CLASSIC AMERICAN STEAMROLLERS

1871 THROUGH 1935
PHOTO ARCHIVE

Judge Raymond L. Drake and
Dr. Robert T. Rhode

Iconografix
Photo Archive Series

Iconografix
PO Box 446
Hudson, Wisconsin 54016 USA

Library of Congress Card Number: 00-135949

ISBN 1-58388-038-0

01 02 03 04 05 06 07 5 4 3 2 1

Printed in the United States of America

Cover and book design by Shawn Glidden

Copy editing by Dylan Frautschi

Book Proposals

Iconografix is a publishing company specializing in books for transportation enthusiasts. We publish in a number of different areas, including Automobiles, Auto Racing, Buses, Construction Equipment, Emergency Equipment, Farming Equipment, Railroads & Trucks. The Iconografix imprint is constantly growing and expanding into new subject areas.

Authors, editors, and knowledgeable enthusiasts in the field of transportation history are invited to contact the Editorial Department at Iconografix, Inc., PO Box 446, Hudson, WI 54016.

ACKNOWLEDGMENTS

Collections Represented in This Book:
California History Section of the California State Library
Clark collection (Dave Clark)
Clark County (Ohio) Historical Society
Crout collection (Verne Crout)
Denver Public Library Western History Collection
Drake collection (Raymond L. Drake)
Erb collection (David T. Erb)
The F. Hal Higgins collection, Department of Special Collections, University of California Library at Davis
Gencor Industries collection
Gillette collection (Beverly Gillette)
Historic Construction Equipment Association (Tom Berry, Archivist and Contributing Editor)
Lantz collection (Jerry Lantz)
Lavender collection (Morris Lavender)
Lawrence Clark collection
Marion County (Ohio) Historical Society
Massillon Museum (Massillon, Ohio)
Owensby collection (Fulton Owensby)
Reynolds Museum (Dan Bodie, Senior Curator, Reynolds-Alberta Museum)
Rhode collection (Dr. Robert T. Rhode)
Roberts collection (Jesse Roberts)
Shanklin collection (Norman Shanklin)
Sourisseau Academy, San Jose State University
Vogel collection (Robert Vogel)
Wittmer collection (Donald Wittmer)
The authors thank road-builders Gary L. Heinold and Sandy Jones-Heinold for their insights; Debbie Pope, Helpdesk Specialist, Office of Information Technology, and Emily Werrell, Coordinator of Instructional Services and Reference Librarian, Main Library, Northern Kentucky University, for their help in preparing this book; Kathryn DeSensi for her valuable research; Morris Lavender of Kewanee, Illinois, who, twenty years ago, went "dumpster diving" to save the factory records of both the Kelly-Springfield and Buffalo-Springfield companies; Jesse Roberts and Roy H. Herr, who ran these machines when they were still in common use; pioneer researcher Jack Norbeck; and Virginia Weygandt, Senior Curator of the Clark County (Ohio) Historical Society, for her help in locating information on Buffalo, Kelly, and Buffalo-Springfield, and William Grimstad, for the cover photograph.

The authors also express their gratitude to the following individuals for help in researching the companies shown in parentheses:

Esther Hays, Beverly Gillette, and Elizabeth Cooley, granddaughter of David Cook (Acme)
Jack Alexander (Atlas, Enright, and Price)
Ellen Harding of the California History Section of the California State Library (Atlas)
Dave Clark (Austin and Iroquois)
Don Bradley, Tom Graham, John Erickson, Mike McKnight, and Derek Rayner (Avery)
Jerry A. Lantz (Bucyrus and Galion)
David Erb (Case)
Robert Hungerford (Coldwell)
Carol Cobbs of the Columbiana Public Library (Enterprise).
Annita Andrick and Stephanie Gaub of the Erie County Historical Society and Museums, and Tom Berry (Erie)
Jean True, Bill Wurts, Fulton Owensby, and Donald Wittmer (Groton)
Gail E. Anderson (Harrisburg)
David F. Brashears, Engineer for Gencor Industries, Suzanne Crowe, Reference Librarian at the Indiana Historical Society, Andrea Bean Hough, Senior Subject Specialist of the Indiana Division of the Indiana State Library, and Matt Hannigan, Reference Librarian in the Business, Science, and Technology Service Section of the Indianapolis-Marion County Public Library (Hetherington & Berner)
Alan C. King, Jane Rupp, Director of the Marion County (Ohio) Historical Society, and Thomas R. Haid (Huber)
Norm Shanklin (Iroquois)
John C. Lindelof, great-grandson of Anders Lindelof, and William Worthington, Curator at the Smithsonian Institute (Lindelof)
T. C. Spires (Marion)
Anne B. Shepherd, Reference Librarian at the Cincinnati Historical Society Library (Ross)
Dan Bodie, Senior Curator of the Reynolds-Alberta Museum (Robert Bell and Waterous)
Mike Hand and John C. Waterous (Waterous)

FOREWORD

by Randy Leffingwell

Sometimes having a good dictionary makes a task much easier. This is especially the case when the proper words are needed to preface a work such as this. This book goes considerably further than what is often called "research," or a "close, careful study," if one accepts what *The American Heritage Dictionary of the English Language* calls it. And this is more than merely a "scholarly or scientific investigation or inquiry."

First of all, the work that Ray Drake and Bob Rhode have done here has resulted in something that is, simply, way too much fun and far too interesting to get buried under so scholarly a classification. Yes, of course, it is scholarly. These two have done their homework. But that's just the point here. They've dug deeply into this subject, and along the way they have integrated into this collection bits and pieces and insights from all the homework that they've ever done. And we readers are the fortunate beneficiaries of this assimilation of photos and words, or more accurately, photos and knowledge.

The second bit of evidence that this is more than just a research project is what is collected and shown here. There are photographs, sure. The book, after all, is called a "photo archive." Again, going to *The American Heritage*, they call an archive, among other things, "a repository for stored memories or information." That's getting closer, because both the text and photos frequently explain just how these wonderful machines were used. And that surely is work that these authors obtained from the memories and words of others.

So that seems to lead to one lasting impression from this book. This reveals depth of research, it provides insightful text, it offers unusual details portrayed in words and pictures. More importantly, it explains the farming contexts out of which these machines grew. And it points out the irony of their evolution: that they came to produce roads that made it easier to get people off the farms and to get farm goods into towns. What these two authors have done is "the systematic study of past human life and culture by the recovery and examination of remaining material evidence...." That, the dictionary says, is archeology.

Seldom has archeology been this entertaining.

Randy Leffingwell
Ojai, California

During America's golden age of road construction, appearances of steamrollers were often cause for great curiosity and excitement. Fortunately, these events were frequently captured by intrepid local photographers, as seen here in this 1918 parade at Denver. Three Iroquois tandems and one Buffalo Pitts three-wheel roller are parading past the Colorado State Capital Building. *Denver Public Library Western History Collection*

INTRODUCTION

Few machines have captured the imagination the way the steamroller has. For generations, children's books and cartoons have featured steamrollers. For a century, toy steamrollers have been treasured gifts and keepsakes. Adults look back on the era of the steamroller as the "good ol' days" and envision rollers chuffing along maple-lined avenues. These machines have added idioms to everyday speech; for example, to be *steamrolled* is to be forced into compliance with a difficult opponent. Steamrollers symbolize progress and strength. Even rollers powered not by steam but by gasoline are popularly called "steamrollers."

Steam-powered road rolling began in America a few years after the Civil War. The first steamrollers bore the Aveling & Porter trademark and were imported from England. At the same time, American inventors like Abbott Q. Ross and Anders Lindelof were experimenting with new steamroller designs. By the 1890s, American-made rollers had come into their own.

Public demand for improved highways gave impetus to the production of steamrollers. In the late 1870s, the sudden popularity of the bicycle spawned cycling clubs that lobbied loudly for smooth road surfaces. The paved streets that appeared in answer to bicyclists' needs were modest projects in comparison to the massive highway construction effort that followed. In 1891, New Jersey became the first state to pass a law providing for state participation in building roads. In turn, New Jersey became the center of the good roads movement. In 1893, the U.S. Department of Agriculture began to evaluate the existing highway system. For Agriculture to oversee such an investigation was only natural since the majority of Americans were involved in farming. Roads provided the means of transporting abundant commodities from rural areas to cities, and, to a limited extent, from cities to farms.

Significantly, 1893 heralded the first American gasoline-powered vehicle, built by Charles Duryea, and the beginning of Rural Free Delivery, which lessened rural isolation and increased the pressure for better roads. In the new century, Henry Ford's mass production techniques enabled millions of Americans to own automobiles. Concurrently, the good roads movement became a crusade.

On July 11, 1916, the quest for improved highways gained sudden momentum when President Woodrow Wilson signed the Federal-Aid Road Act. All states were forced to adopt tight control before accepting federal advice and financial support in building roads. In that same year, the Kelly-Springfield Company of Springfield, Ohio, and the Buffalo Steam Roller Company of Buffalo, New York, merged. Marking the beginning of the golden age of road building, this marriage resulted in the creation of the Buffalo-Springfield Company, which became the giant of the compaction industry. With Buffalo-Springfield leading the way, America was able to embark on a massive highway construction campaign during the 1920s.

While American fascination with agricultural engines, stationary steam engines, and steam locomotives has continued unabated since the steam era, steamroller restoration and research into road roller history have only recently begun to assume their rightful place in the steam hobby. Not long ago, co-author Raymond L. Drake spent several dirty but delightful hours piecing together parts of an exceedingly rare set of original black canvas curtains belonging to a Buffalo-Springfield steamroller. He was searching not only for details of their construction but also for the printing on them. Like an archaeologist, Drake painstakingly lifted layers of grime to reveal bone-white stenciled lettering, which read, "Buffalo-Springfield Steam Roller Company." Said Drake, "It was like piecing together the Dead Sea scrolls." As Buffalo-Springfield owner Howard Gorin put it, Drake was likely the first person to view such lettering in a generation. Now it remains for Drake to assemble new curtains matching the originals and to make his findings available to other preservationists. This is an example of a trend among roller enthusiasts to recreate components and materials that enable fellow hobbyists to restore their machines accurately.

When the authors of this book began their work, they set the modest goal of publishing original photographs of Buffalo and Kelly-Springfield steamrollers. At the suggestion of Dylan Frautschi, editor at Iconografix, the authors were encouraged to expand the scope of their study to cover all makes of American rollers powered by steam. Early in the project, hobbyists and researchers across North America generously threw open their collections and archives, revealing a treasure-trove of material. The writers discovered that they could offer an in-depth look at the history of American steam road roller production. What has emerged in this publication is the first complete picture of steamroller manufacturing in North America.

Raymond L. Drake
Robert T. Rhode
October 25, 2000

A right-hand view of Acme steam road roller. This roller closely resembles the Buffalo models first introduced in 1893, many features of which were copied by competitors. The most notable differences between Acme and Buffalo are the Acme's long canopy, different gearing, and presence of a steamdome. *Drake collection*

Acme Road Machinery Company, Frankfort, New York

Brothers Walter and David Cook began the Acme Road Machinery Company. During the 1890s, they started their careers as salesmen with the Climax Company of Marathon, New York. Climax was engaged in the manufacturing of rock crushers, gravel machines, and horse-drawn graders. The Cook brothers foresaw the need for the steamroller. In 1898, they founded their own firm in Frankfort. Their Acme Company started successfully with the 1900 introduction of a three-wheel steamroller offered in 10-, 12-, and 15-ton sizes. The firm soon added center dump wagons, crushers, and road graders to its catalog. The Acme Company expanded and erected substantial buildings, some of which are in use at the time of this writing. During the 1920s, Acme offered a diverse line of gasoline-powered rollers and, as late as 1928, was still marketing steamrollers. Acme continued to sell gasoline rollers until the company's dissolution in 1965.

A large part of the Acme plant devoted itself to the manufacture of rock crushers. In this 1923 photo can be seen one of the company's stone crushers, said to weigh 95,000 pounds. Large crushers like this one supplied the much-needed gravel for road paving. *Gillette collection*

A left-hand view of Acme roller equipped with a beltwheel that could be utilized to run crushers or gravel plants. The purpose of the tall pipe attached to the dome is to release steam from the pop-off valve harmlessly away from the engineer. *Drake collection*

This picture clearly shows the double-cylinder scarifier that was unique to Acme steamrollers. While Acme offered 10-, 12-, and 15-ton rollers, it would appear that the only difference between these machines was the addition of wider and heavier wheels or rolls on the larger sizes. *Drake collection*

The Atlas Iron Works, San Francisco, California

Accompanying an article entitled "Building Roads and Pavements" in *The Mining and Scientific Press* for April 14, 1888, was a cut of the Atlas Iron Works tandem steamroller manufactured in San Francisco, California. Atlas rollers ranged in size from 3 to 20 tons and were used in San Diego, Los Angeles, Santa Barbara, San Jose, and Santa Cruz.

This San Francisco built behemoth was the property of the Santa Cruz Pavement Company. It has been said that this firm built rollers that weighed up to twenty tons. This surely is one of the company's largest machines. Like virtually all early tandem rollers (Erie, Iroquois, and Monarch, to name a few), this machine bears a strong resemblance to the Lindelof design. *California History Section of the California State Library*

F. C. Austin Manufacturing Company and The Austin Manufacturing Company, Harvey, Illinois

In 1888, the F. C. Austin Manufacturing Company was incorporated. Shortly after the turn of the century, when the same interests involved in the Western Wheeled Scraper Company of Mount Pleasant, Iowa, purchased the firm, the name changed to the Austin Manufacturing Company. Austin was a manufacturer of scrapers and graders, primarily built in Aurora, Illinois. In the mid-1920s, the company adapted Fordson tractors to its scrapers. Austin offered steamrollers in 10- and 12-ton sizes, but Austin's gasoline-powered rollers eclipsed their steamrollers. Austin designed a large two-cylinder opposed gasoline roller, of which several examples still exist. In the last half of the twentieth century, Austin's six-wheel-drive graders were used throughout America.

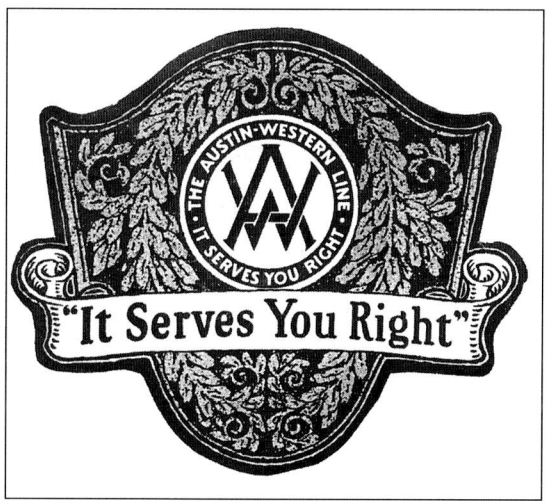

This left rear view of the Austin roller shows the scarifier and the engineer's compartment clearly. Austin was one of the only companies to place the firm's name around the flywheel and at the front of the smokestack. These machines were offered in 10- and 12-ton sizes. By 1928, Austin was out of the steamroller business. *Drake collection*

The Avery Company, Peoria, Illinois

The story of the Avery Company begins in the Andersonville Confederate Prison during the Civil War. A captured Union soldier named Robert H. Avery sketched a design for a corn planter in the soil of the prison yard. By 1874, the Kansas native had assembled a working model. In 1877, Robert and his brother Cyrus M. Avery created a company in their name in Galesburg, Illinois. Better transportation facilities lured the brothers to Peoria, Illinois. In 1883, the firm's name became the Avery Planter Company. It was not until 1891 that Avery began to manufacture steam engines and yellow wood threshers known as Yellow Fellows. Robert died in 1892, and Cyrus took over the presidency, naming the legendary J. B. Bartholomew vice-president. At the turn of the century, the firm reorganized as the Avery Manufacturing Company. Cyrus died in 1905, and Bartholomew became president, carrying on the expansion that had begun under the Avery brothers and leading the renamed Avery Company to ambitious heights.

The first Avery engines were of a top-mounted return-flue style. The company also built a top-mounted locomotive type of engine. The double undermounted engine became the Avery steamer of preference. On September 19, 1903, Martin J. Hogan of Canton, Ohio, received a patent for the undermount design. He probably was an employee of the C. Aultman Company, which built undermounted engines. The 20th Century undermounted steam engines, manufactured in Boynton, Pennsylvania, were licensed under the Hogan patent. In 1906, the patent was sold to Avery. Bartholomew once won a case against the A. W. Stevens Company of Marinette, Wisconsin, for infringement of the Avery patented undermount design.

The company reached its pinnacle about 1920. The farm depression that started in 1921 sent Avery into a decline. Bankruptcy and receivership came in early 1924, and Bartholomew died in 1925.

The Avery undermounted road roller consisted of a conversion package that transformed the agricultural engine into a roller by replacing the front truck with a special roller front truck and by removing the lugs from the driver wheels.

Avery was another farm traction engine company that elected to enter into the road rolling business through the modification of one of the firm's agricultural engines. The illustration to the right was featured in the company's catalogs from 1910 through 1916. It is doubtful if many of these machines were sold. The Avery undermounted engine resembles a railroad locomotive. Avery was one of only four companies to build such undermounted steamers. The famous trademark for Avery machines was a very toothy dog, and Avery products were universally known as "The Bull Dog Line."

Baker

The A.D. Baker Company, Swanton, Ohio

In 1898, Abner D. Baker built his first agricultural steam traction engine. Between 1898 and 1927, his firm manufactured approximately 1,800 farm engines. Baker is perhaps best known for his valve gear, which he adapted for locomotives, leading to the creation of the Pilliod Company. Baker began producing threshing machines in 1908. The company added steamrollers to its product list in 1909. Baker rollers had two types of front roller and steering mechanism, but all weighed ten tons. Between 1909 and 1927, Baker sold 157 road rollers. The following figures show Baker's steamroller output: 1909=1 roller, 1910=13 rollers, 1911=16 rollers, 1912=8 rollers, 1913=13 rollers, 1914=22 rollers, 1915=31 rollers, 1916=33 rollers, 1917=9 rollers, 1918=6 rollers, 1919=1 roller, 1926=2 rollers, 1927=2 rollers. These figures are from the A. D. Baker Company Traction Engine Book, courtesy of Louis Abner Carson, Baker's great-grandson, Vice-President of Scottdel, Inc.

Baker and Henry Ford were friends, and Ford was a frequent visitor at Baker's house. According to Rosabelle Krauss, Baker's former housekeeper, Ford and Baker enjoyed discussing inventions—especially those that fizzled.

Like other builders of farm steam traction engines, the A. D. Baker Company chose to modify its engine so that it became a road roller. This left-hand view clearly shows the roller's agricultural engine origins. *Baker catalog*

In this right-hand view, the unusual saddle support for the front steering yoke that was used on early Baker road rollers may be seen. The Baker rollers were 10-ton rated machines. These rollers could be steered either by hand or with friction assist. *Baker catalog*

On this model, the front steering yoke appears to be similar to that employed on Case rollers. The first Baker roller was built in 1909, and a total of 157 were manufactured. *Baker catalog*

Birdsall

Birdsall Engine Company, Auburn, New York

In 1860 in Penn Yan, New York, Hiram Birdsall and Edgar M. Birdsall began H. Birdsall & Son. By 1881, the firm had outgrown its quarters, moving to the Cayuga Chief Company's former plant in Auburn. Birdsall produced its first traction engine after the move. Across its history, the Birdsall Engine Company built such machines as threshers, horse powers, traction and portable engines, sawmills, mowers, and other agricultural implements. The Birdsall used what later was called an "automobile steer," even though the apparatus was made in the 1880s before the automobile. Elmer Ritzman, founding editor of *The Iron-Men Album Magazine*, reported that early automobile manufacturers paid Birdsall a royalty for the steering patent. The Birdsall Engine Company invented a zigzag pattern of drive wheel face known as the "lattice wheel," designed to cut through soft ground. Birdsall also built a solid drive wheel face, used on its steamrollers. The steering wheel of the road rollers was located on the engine's right-hand side, but the steering wheels of the company's other engines were on the left-hand side. Birdsall sold a "heavy roller" engine that probably served as a road locomotive since it had two front wheels rather than a front roller.

Here is another example of a manufacturer of farm traction engines deciding that, with the addition of a front roll, an agricultural engine can be transformed into a steamroller. That the Birdsall roller is extremely rare and virtually unknown, even to experts on this marque, suggests that such ventures were not always crowned with success. T. H. Smith's *The Album of American Steam Traction Engines* (1953)

The Bucyrus Road Machinery Company, Bucyrus, Ohio

The Bucyrus Road Machinery Company was a late arrival to the industry. The firm first appeared in 1923. The opening page of the 1923 catalog advertised Bucyrus steam and motor rollers—macadam, tandem, combination, golf, asphalt, maintenance, embankment, and special rollers for all purposes—road scarifiers, and general roadbuilding machinery. The catalog claimed that the design of the Bucyrus roller capitalized on twenty-five years of experience. A dry pipe inside the boiler ensured that hills would not interfere with the delivery of dry steam to the engine. The boiler carried 200 pounds per square inch. It is probable that the firm manufactured steamrollers from 1923 until 1928.

The Bucyrus scarifier was pictured in the company's 1923 catalog. *Lantz Collection*

This front shot of a Bucyrus roller could easily be mistaken for a Galion or even a Buffalo-Springfield. A late arrival (1923) to the compaction industry, these machines clearly are copies of roller designs built by others. *Wittmer collection*

The front roll, yoke, and kingpin housing on the Bucyrus roller appear to be virtually identical to those found on the Buffalo-Springfield, while the rear wheels closely resemble those used on Galion products. *Wittmer collection*

Buffalo Pitts

The Buffalo Pitts Company and The Buffalo Steam Roller Company, Buffalo, New York

In 1799, the nearly legendary twin brothers John A. and Hiram A. Pitts were born in Winthrop, Maine. They first built tread powers to run groundhog threshers, so named because the wheat was thrust down a chute reminiscent of a groundhog's hole. Beginning in 1830 and 1831, the brothers experimented with designs of threshers that would clean the grain. In 1837, they patented their apron-conveyor thresher. Numerous early threshers were modeled on it. By 1847, the Pitts brothers were selling threshers in Illinois. John Pitts left to market threshers in Ohio. In 1851, Hiram A. sold Chicago Pitts threshers in Illinois. Meanwhile, John traveled to Buffalo, New York, where he produced Buffalo Pitts threshers in 1851. The firm of Pitts & Brayley existed in 1859; John died that year. In 1860, the company sold portable steam engines. By 1866, the firm was known as the Brayley and Pitts Works, named for John B. Pitts, the elder John's son, and James Brayley, the elder John's son-in-law. In 1877, the Pitts Agricultural Works was incorporated. After 1897, the firm was called the Buffalo Pitts Company.

The company perfected its roller design and produced the first Buffalo Pitts Niagara Road Roller by 1892, later forming the offshoot business called the Buffalo Steam Roller Company. By 1901, the firm had already sold several of the Buffalo Pitts Double Engine Steam Road Rollers. In 1910, Marquis J. Todd assigned to the Buffalo Steam Roller Company his patent for a front roll with a removable center section, converting the three-wheel rollers into a road locomotive. The company ceased to exist in 1935.

This is the earliest known image of a Buffalo steamroller. These machines were called Niagara models and were obviously an attempt to convert a farm traction engine into a road roller. It is interesting just how closely the front kingpin housing resembles that used on Kelly rollers and that these machines had steamdomes. This ca. 1892 roller also had a double-cylinder engine. *Drake collection*

Buffalo Pitts 10-ton steamroller built in 1893. This is the first Buffalo Pitts roller designed as a steamroller—not merely a reworked farm traction engine. In this 1925 picture, the roller was still in regular service. *Drake collection*

This beautifully detailed cut shows a ca. 1900 Buffalo Pitts steamroller. The heavy front steering yoke indicates that this is a larger size, probably a 12-ton machine. The Buffalo trademark on the coal bunker was used on all the firm's rollers prior to the 1916 merger with Kelly-Springfield. *Drake collection*

This 1909 picture shows the Buffalo Pitts Improved Scarifier being pulled by one of that firm's steamrollers. The hand cranks enable the engineer to adjust the teeth of the scarifier to a uniform depth. *Drake collection*

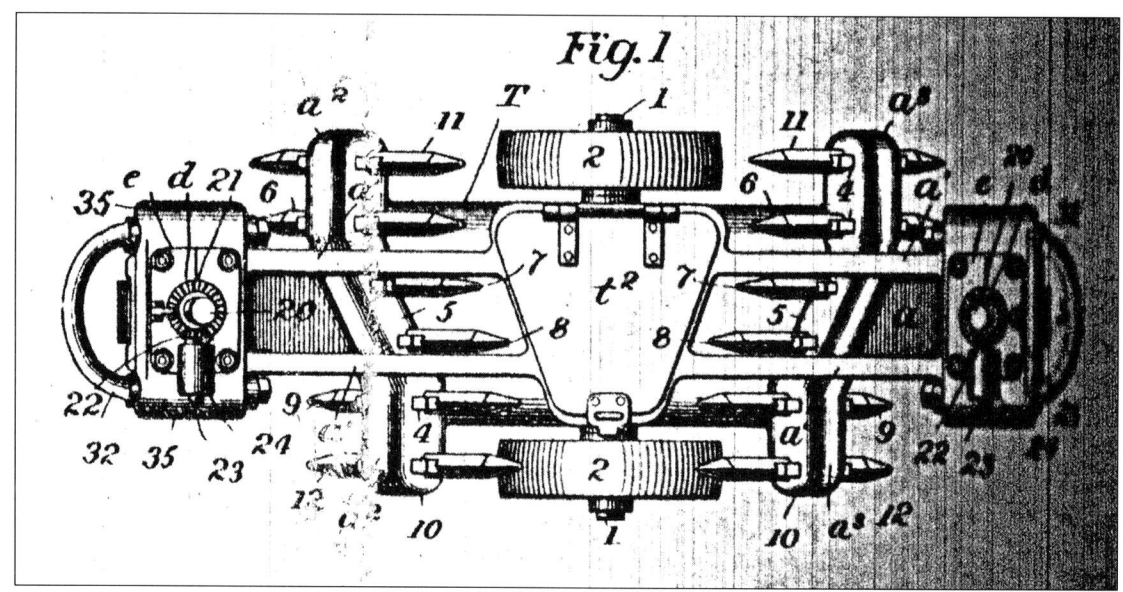

Here is the original patent drawing for the Buffalo Pitts Improved Scarifier. The date of this patent is August 27, 1907.

This picture shows the first type of scarifier to be mounted on a steamroller. This invention was introduced by Buffalo Pitts in 1912 and was manually operated using the hand pulleys that can be seen on the side and at the rear of the machine. Such side-mounted units are called "berm" scarifiers. This type of roller was known as a Special Maintenance roller. Such rollers bore the designation "SM." If this machine was not equipped with the berm scarifier, it would be called a "Steam Macadam" type roller—with the designation SM standing for Steam Macadam. *Drake collection*

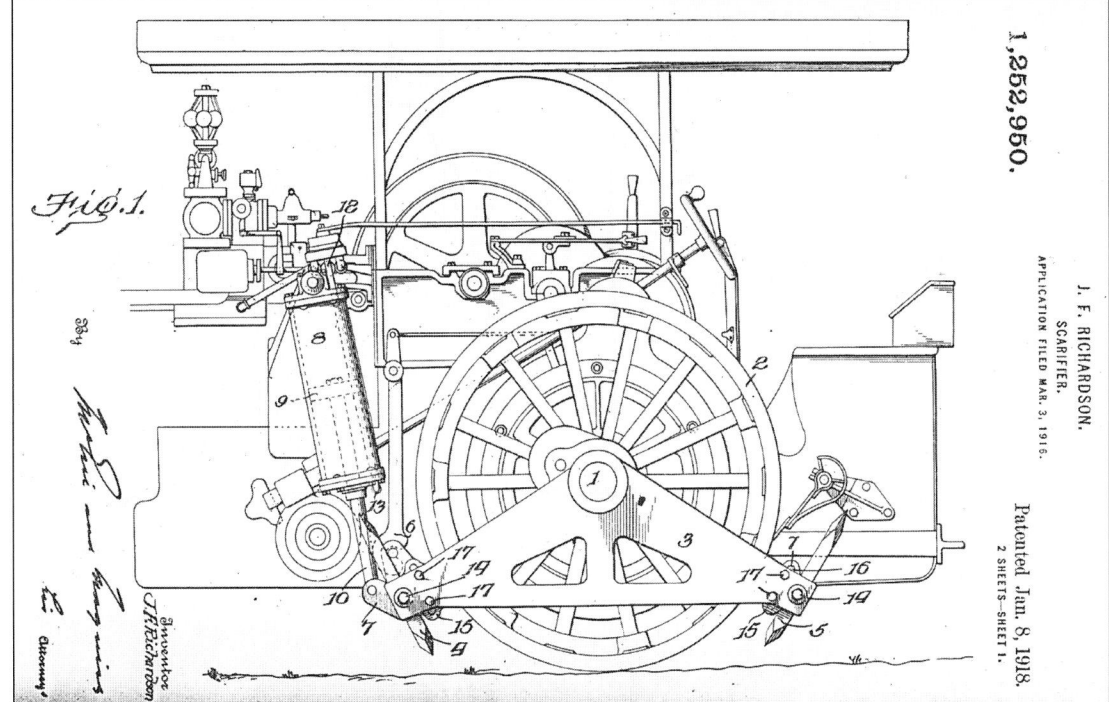

It did not take long for someone to figure out that a scarifier would work considerably better if it were steam powered. That someone was Jeffers F. Richardson of Buffalo, New York, and it was he who designed a steam cylinder arrangement that greatly improved the berm scarifier. He assigned his invention to the Buffalo Steam Roller Company. While the patent date is January 8, 1918, there is no doubt that these scarifiers were in production a year prior to that date.

Seen here is an early 12-ton SM44 "Special Maintenance" roller, built in February of 1918. There is no evidence to suggest that this type of scarifier was ever used on the SM36 10-ton rollers. These early scarifiers did not have the 6" x 8" brass instruction plaque attached to the steam cylinder. That practice appears to have started a few months later. The Dietz headlamp was used from 1912 until the end of production and is a "Champion" model. *Drake collection*

Here may be seen a Buffalo Pitts "Special Tractor" hauling a train of gravel-filled cars. Period literature stated that this type of train could haul forty tons of rock with relative ease. The "elephant trunk," or rigid coupler, was used to pull or push the wagons or cars. *Drake collection*

This picture shows the very rare rigid front coupler that could be ordered on "Standard Macadam" and "Special Tractor" rollers. In this view may be seen the coupler attached to a "Special Tractor" or road locomotive roller specially designed for highway use. *Drake collection*

This is a 10-ton model SM36 roller named an "Extremist" by the factory. Of note are the cinder screen on the smokestack and the rack attached at the rear. The rack indicates that this was a wood-burning machine. This roller was built at Buffalo in March of 1920 and is serial number 9502, making it one of the later machines to be built at the New York factory. The curved corner brackets on the canopy are features found only on rollers made in New York. Also noteworthy is the early square-shaped headlamp seen on the top of the kingpin housing. *Drake collection*

This 1916 picture shows one of the first, if not the first, 10-ton SM36 steamrollers to be fitted with a rear-mounted Buffalo Pitts scarifier. This extremely clear photo gives a fine view of the pin striping and lettering design. *Drake collection*

This is another view of the machine seen on the previous page. The angled bracket on the top of the kingpin cap is the mounting bracket for the Dietz brand "Champion" headlamp used from about 1912 until the end of production. *Drake collection*

These two photographs show an unusual pulley that is mounted on the rear wheel of an early Buffalo Pitts "Special Tractor." According to old-time roller man Roy Herr, such pulleys were useful in extracting a mired roller from the mud through the stratagem of looping a rope around the pulley and having the opposite end of the rope attached to a very large tree. *Crout collection*

27

This photo demonstrates just how a Buffalo (New York built) steamroller could pull a road grader that normally would be drawn by horses. This picture, showing an early wooden-wheeled grader, was taken shortly after 1900. *Clark County (Ohio) Historical Society*

(top left) The first Buffalo Pitts tandem steamroller introduced in 1892 was strikingly similar to the Anders Lindelof roller of 1873, a drawing of which can be seen in the upper left of this picture. The primary difference between these two machines is that the Buffalo had a double set of beveled gears around the main roll, whereas the Lindelof roller used a single set.

(top right) About 1906, Buffalo moved the engine between the boiler and rear wheel and installed double spur gears that had been patented by the O. S. Kelly Company in 1902.

(center) Around 1910, Buffalo adopted another Kelly improvement—the horizontal engine—which then made it fairly difficult to distinguish between the two makes.

(bottom right) A front view of a Buffalo tandem roller. While the identification plaque on a Kelly is located on the left side of the "gooseneck" and, after 1916, was in the same location on Buffalo-Springfields, the Buffalo Pitts rollers have their identification plaques mounted at the front of the gooseneck around the area that holds the kingpin. *All from Drake collection*

In this picture, taken about 1907, a crew of five men work with a Buffalo (New York built) tandem steamroller and farm plow to rip up an old roadbed. While a tedious process, it was easier than employing horses. The style of roller, with its upright (not slanted) engines and enclosed gears, indicates that this roller is not newer than 1910 nor older than 1906. *Clark County (Ohio) Historical Society*

In the early 1900s, the Buffalo Steam Roller Company, like their competitors in the compaction industry, the O. S. Kelly Company of Springfield, Ohio, elected to enter the truck manufacturing business. Here may be seen the Buffalo steam-powered dump truck. Close examination of the identification plaque reveals it to be the same as that used on the firm's steamrollers. Like the Kelly Company, Buffalo made only a few steam-powered trucks. *Drake collection*

Buffalo-Springfield

Buffalo-Springfield, Springfield, Ohio

This is the most widely recognized name of all American road rollers. The roots of the company reach far back. By the 1880s O. S. Kelly in Springfield, Ohio, and the Pitts Agricultural Works in Buffalo, New York, were building agricultural engines. In the 1890s both companies recognized the need for a reliable steamroller. Their first rollers were simply their traction engines with smooth rear wheels and a solid roll in place of the front wheels. By 1892 both Kelly-Springfield and Buffalo were marketing steamrollers. During the next twenty-five years, both firms were innovative. In 1907 and 1908, Kelly built the first gasoline-powered rollers in America. The industry failed to recognize this milestone, and it was not until April of 1910 that the first order was received. This machine was serial number 2281, sold to the city of Minneapolis, Minnesota. In 1912 Buffalo Pitts brought forward another important invention, the scarifier. The first ones were mounted on the left side of the roller and were called "berm" scarifiers. By 1915 this was greatly improved with the addition of a steam cylinder, and in 1916 a rear-mounted steam scarifier was offered. The berm scarifier ripped a two-feet-wide area, the rear-mounted one a five-feet-wide area—a marked improvement.

The early twentieth century witnessed two events with far-reaching implications for roads. The first took place in 1913 when the federal government announced an ambitious plan to construct over fifty thousand miles of highways. Second, in 1916, the largest manufacturers of road rollers, Kelly-Springfield and the Buffalo Steam Roller Company merged to become the premier manufacturer of American road rollers. The arch-rivals recognized a profitable opportunity in the sweeping changes taking place

under the Federal-Aid Road Act of that year. They wanted to corner the steamroller market. Manufacturing of the combined companies was to be centered in Springfield. While the merged firms intended to name their rollers "Buffalo-Springfield," a Buffalo stockholder refused to endorse the transfer agreement. Company stationery from 1917 reflects that, thanks to the stockholder, the firm had to call itself the "Buffalo-Springfield Company, exclusive manufacturers of Buffalo Pitts steamrollers and Kelly-Springfield steamrollers." Kelly-Springfield rollers and Buffalo Pitts rollers were manufactured under the same roof in Springfield. Also, Buffalo Pitts rollers were built in Buffalo. The New York rollers had curved canopy corner brackets, the Springfield rollers straight brackets at 45-degree angles. The New York rollers boasted a lighter steering yoke; Springfield ones used the 10-ton Kelly yoke. Also, the New York rollers had an extra bolt in the corner of the large kingpin housing.

The firms had courted each other prior to the merger. In fact, the first steamroller to be manufactured at the Springfield plant using Buffalo Pitts components was serial number 2724, built on January 18, 1913—three years before the official merger! The first roller built at the Springfield factory that bore the Buffalo name was a five-ton tandem, serial number 3422, finished on March 25, 1916. After the merger, the Springfield factory expanded, and manufacturing at the Buffalo plant began to decline. In 1921, production there ceased altogether. In that same year, the stockholder relented, permitting the name "Buffalo-Springfield" to appear on all models of roller built by the firm. Certain batches of rollers were completed in 1921 and called Buffalo-Springfields, even though sister engines from the same batch but completed earlier were named Buffalo Pitts rollers.

This picture shows a very early Buffalo-Springfield SM36 (a model designation, standing for "Steam Macadam") 10-ton roller. There are at least three of these "transitional" rollers still in existence. While this machine was built at Springfield, as evidenced by the squared yoke, and the angled canopy brackets, it more or less retains certain Buffalo Pitts features, such as the toolbox mounted on the right side and the kingpin housing that has the extra bolt in the right corner. Also retained is the large floral design on the yoke. *Drake collection*

Shown here is a model SM44 or 12-ton (actual operating weight 32,600 pounds) steamroller. While the toolbox proclaims "Buffalo Pitts," this machine, built in February of 1918, was a product of the Springfield factory. The easiest way to identify a 12- or 15-ton roller is that they both have a solid area, or a counterweight, between two of the spokes on the flywheel. On the 10-ton Buffalo rollers, the area between the spokes is open. The factory name for the 12-ton machine was "Hospondar," which means Servant of the Gods. *Drake collection*

The left-side view of the SM44 12-ton roller seen on the previous page. In photographs, various models of Buffalo-Springfield rollers may appear to be the same size, but, in reality, they are progressively larger. For example, a 10-ton roller weighs 27,750 pounds, is 208" long, and is 80" wide, with standard 18" rear wheels, whereas a 12-ton roller weighs 32,605 pounds, is 222" long, and is 89" wide. A 15-ton roller weighs 38,640 pounds, is 238" long, and is 95" wide. *Crout collection*

Introduced in January of 1925, this is the SM36 "Extolled" 10-ton roller. The first three-speed models began with serial number 12741 and were used only on 10-ton (25,130 pounds empty, 27,750 pounds loaded weight) machines. The vast majority of Buffalo rollers, except for a few very early single-speed models, had two-speed transmissions. The idea of a third gear was to give the roller a "high gear" to move from job site to job site at a speed up to ten miles an hour. Raymond L. Drake, one of the co-authors of this book, opened his roller (serial number 12760) once in third gear and does not plan to do so again, explaining that, since steamrollers are not equipped with brakes, an emergency stop could be problematical. *Drake collection*

A close-up and highly detailed view of a Buffalo-Springfield double-cylinder engine used on all three-wheel steamrollers. While there are differences between the 10-, 12-, and 15-ton bore and stroke sizes, the engines are virtually identical visually and remained so from the early 1900s until the end of production. All engines used a Stephenson valve gear with a Pickering governor and a Manzel lubricator. This machine is a 1925 or later model, as it has a three-speed transmission and a side-mounted sight glass. *Clark County (Ohio) Historical Society*

In this picture may be seen ten Buffalo-Springfield 10-ton SM36 steamrollers loaded on railroad flatcars in front of the factory at Springfield. Close examination of this load shows how rollers should always be secured for transport. Only the rear wheels are to be blocked front and rear, as well as on the sides, while the front wheel is to be blocked only on the sides, never on the front and rear. If the roller were to move, serious damage to the steering fork could result from blocking the front wheel front and rear. The lettering design and black painted machines indicate that this photo was taken in 1921. The flatcars are from a mix of railroads, the first from the Atlantic Central line, the second from the Pennsylvania, and the third from the Southern Indiana. *Clark County (Ohio) Historical Society*

Here is a picture of an SM44 12-ton roller equipped with a berm, or side-mounted, scarifier. This type of scarifier was never offered on the 10-ton SM36 rollers. Of note is the serial number plaque (above the builder's plate) which states that this roller is serial number 4855, indicating that it was built on July 17, 1921, making this one of the last to be produced in the early serial number range. There has been confusion surrounding Buffalo serial numbers, and the following explanation may help to clear up the mystery. After the 1916 merger of Buffalo Pitts and Kelly Springfield, rollers were built at both the Springfield and Buffalo plants until April of 1921, when production ceased at the Buffalo facility. During this period, each factory used a separate numbering system. The last roller produced at Buffalo was serial number 10359, and the last of the early range numbers produced at Springfield was serial number 4863. On April 30, 1921, the new numbering system started at serial number 10360. Certain batches of rollers were in production before April 30th and the serial numbers within those batches had been predetermined; therefore, even certain rollers built after April 30 retained the serial numbers of the earlier batches and did not receive serial numbers from the new system. This practice is common in manufacturing and leads to serial number overlap. *Drake collection*

Here is another picture of a 12-ton SM44 steamroller that is also equipped with a berm scarifier. This picture was taken about 1922 and shows the scarifier set in the forward position, whereas the previous picture showed it set in position for backing up. This roller has the 6" x 8" instruction plate on the steam cylinder and the Buffalo-Springfield decal on the belly tank. *Clark County (Ohio) Historical Society*

This close up shot of the rear wheel and scarifier of the roller seen on this page not only shows the wide gold stripes but also clearly details the fine accent striping that went into these machines. This sort of attention to detail demonstrates the pride of workmanship that the employees put into these rollers.

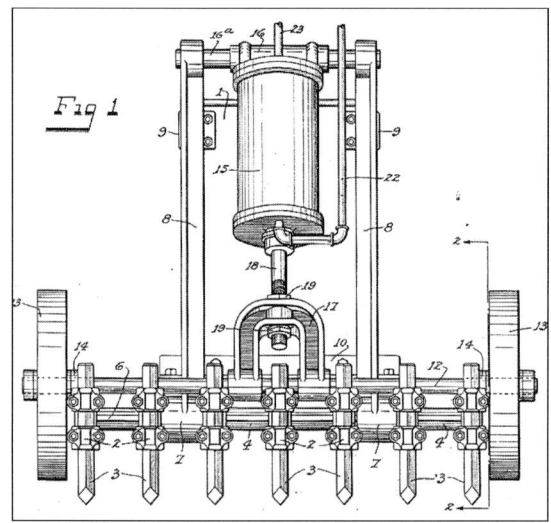

On June 23, 1915, Henry D. Jones of Springfield filed for a patent on a new type of pressure scarifier. For the next quarter of a century, his invention was widely used on Buffalo-Springfield steam- and gasoline-powered rollers and copied by many of that company's competitors. Here are the original patent papers, approved on May 15, 1917. The inventor assigned the patent to the Kelly-Springfield Company; however, it was always called a Buffalo-Springfield Pressure Scarifier, even when installed on Kelly-Springfield rollers.

This 1916 photograph, taken in Westchester County, New York, is the earliest picture known of a Buffalo-Springfield Pressure Scarifier. The side wheels of the scarifier have squared edges like those in the patent drawings and unlike those with rounded edges seen on all other rollers. The steam cylinder has "Patent Pending" stenciled on it, while regular production models have an instruction plate attached. *Drake collection*

In this 1919 photograph, an early scarifier can be seen that has the standard rounded-edge wheels and also the brass instruction plate attached to the steam cylinder. This scarifier has the rarely seen ribbed steam cylinder. It appears that this type of cylinder was used on some 1918-1919 rollers and was discontinued after those years in favor of the smooth cylinder. At least one of these early ribbed-cylinder machines still exists. *Drake collection*

This picture, taken in February of 1918, shows a Buffalo-Springfield steamroller that has the smooth steam cylinder used up through the 1940s. What makes this shot unusual is that the roller is fitted with a scarifier having thirteen teeth instead of the customary six. This was possibly an experimental design, since this is the only picture showing this arrangement that is known to exist. Also of note is that this roller is fitted with the elephant trunk rigid front coupler. Apart from the gray metal jacket, this 10-ton roller appears to be painted black. *Drake collection*

In this picture may be seen a 7-ton model ST#21 (with ST standing for "Steam Tandem"), which has been extensively retouched by the factory art department. The 7-ton model was given the name "Conqueror." This size machine was the most popular of the tandem rollers; consequently, 7-ton steamrollers are the most commonly found survivors today. It is interesting to note that this roller is equipped with the horizontally mounted engine that was used through the end of production. *Drake collection*

Here is a most unusual tandem roller. This design, with its upright side-mounted double-cylinder engine, was commonly found on most early tandem rollers. The upright engine design caused the roller to rock, which created a wavy pattern in the pavement. It was for this reason that both Kelly and Buffalo abandoned this style in favor of the horizontally mounted engine after 1910. This machine, serial number 13594, a 7-ton tandem, from job #275, was built on February 5, 1927, long after this style was obsolete! It was probably a special-order roller. *Clark County (Ohio) Historical Society*

This picture shows the right side of tandem steamroller serial number 13594 and clearly shows the factory lettering and pin-striping designs. The model designation for these machines was ST#21. *Clark County (Ohio) Historical Society*

This picture shows six Buffalo-Springfield 10-ton tandem steamrollers that were shipped to the Highway Construction Company of Miami, Florida, in 1921. It was at this time that Carl Fisher of Detroit, Michigan, was converting a mosquito-infested swamp into what we now call Miami Beach. There are two interesting points regarding these machines; first, they are painted black, which all Buffalo rollers were until late 1921, when they began using a Brewster green color, and, second, the cylindrical drums in the coal trays indicate that these were oil-fired rollers. For steamrollers that were to be used in places where coal was not readily available, wood- or oil-fired conversion rollers could be special-ordered. Also of note is just how large these rollers look compared to the factory employee at the left side of this photograph. *Drake collection*

Like an automobile dealership, a steam-roller dealer needed to have a parts department. In this picture, the inventory of the Spears-Wells parts department can be seen clearly. It would be the dream of a lifetime for one of today's restorers to locate a cache like the one pictured here. *Clark County (Ohio) Historical Society*

Just like automobile manufacturers who had dealer networks to distribute their products, steamroller companies had dealers located in most major cities. Here is a view of one of those dealers, the Spears-Wells Machinery Company, which has two rollers on display just inside the front door. *Clark County (Ohio) Historical Society*

By the early 1930s, many steamrollers were over a quarter of a century old and were in need of repairs or replacement. Often a community's water was so acidic that it destroyed a boiler in just a few years. If in the 1930s a town's steamroller gave out, the community's leaders faced a serious dilemma. While, on the one hand, most citizens demanded government services, including good roads, on the other hand, with millions unemployed, people standing in soup lines, and the air rife with talk of Red revolution, public officials could hardly justify spending thousands of dollars on a new road roller. The engineering department of the Buffalo-Springfield company came up with a compromise: convert rollers from steam to gas! Here is an 8-ton roller, serial number 10392, originally built on April 26, 1921, converted to a gasoline engine on August 25, 1934. Worthy of note is that this roller has the subdued striping design common among mid-1930s rollers and that it has a brake on the rear wheel. *Drake collection*

While converting steam-powered tandem rollers was relatively easy, converting a three-wheel roller was an entirely different matter. Leave it to the geniuses at the Springfield plant to come up with a solution! Simply remove the engine from the boiler, then replace it with a Waukesha gas engine and transmission, and—presto!—you now have a new (did we say new?) road roller for a few hundred dollars. Here may be seen a 1919 Buffalo Pitts, serial number 9293, that underwent this metamorphosis on March 29, 1930. Old-time roller engineers, such as Jesse Roberts of New York, said that, while these conversions worked reasonably well, they did take forever to back up. Surprisingly, several of these conversions still exist. All this goes to prove that Rube Goldberg was alive and well in the Springfield engineering department during the Great Depression. *Drake collection*

In the above picture is the last steamroller to be built in America. This machine was serial number 17490. It was a Buffalo-Springfield 7-ton Conqueror model, and it was photographed on February 2, 1935, the date that it was completed at the Springfield factory. An easy way to identify the late production steamrollers is by the subdued pin-stripe pattern. For example, it is of note that the flower pattern is absent from the steering yoke, as is common on all rollers made after 1933. This machine marked the end of an era for the road roller industry, when steam had long since ceased to be king. *Clark County (Ohio) Historical Society*

The J. I. Case Threshing Machine Company, Racine, Wisconsin

The industry giant in the manufacture of agricultural steam engines was the J. I. Case Threshing Machine Company. Serial numbers for Case engines reached 35838 in 1926, when the last Case steamer was built. The success story began in 1842, when Jerome Increase Case bought six groundhog threshers. He left Oswego, New York, traveled by steamship to Chicago, and journeyed to Rochester, Wisconsin. Between Chicago and Rochester, he sold five of the threshers. With the sixth, he did custom threshing. By 1844, Case became a manufacturer. Since Case could not secure rights to build a millrace in Rochester, he moved to Racine to establish his factory. His first threshers were made under license from the Pitts brothers and from Jacob V. A. Wemple, a partner of George Westinghouse who patented a threshing mechanism similar to the Pitts apron. The Case firm began a virtually uninterrupted growth cycle. Good workmanship, intelligent designs, and easy credit created the Case hegemony.

The first ten-ton Case steamroller was serial number 15901, built in 1905. The first twelve-ton roller appeared as serial number 20649 in 1908. The former size proved more popular, as shown by the following statistics: 1905=9 ten-ton steamrollers, 1906=40 ten-ton rollers, 1907=25 ten-ton rollers, 1908=75 ten-ton rollers and 3 twelve-ton rollers, 1909=7 twelve-ton rollers, 1910=146 ten-ton rollers, 1911=126 ten-ton rollers, 1912=none, 1913=none, 1914=50 ten-ton rollers, 1915=51 ten-ton rollers, 1916=none, 1917=25 ten-ton rollers, 1918=4 ten-ton rollers, 1919=46 ten-ton rollers, 1920=17 ten-ton rollers and 8 twelve-ton rollers,

1921=40 ten-ton rollers and 10 twelve-ton rollers, 1922=none, 1923=24 ten-ton rollers and 1 twelve-ton roller. According to these figures, Case built 678 ten-ton steamrollers and 29 twelve-ton rollers, for a total of 707 steamrollers.

Case literature said that steamrollers could have tandem compound engines. Power steering was an option. Case rollers could be converted to traction engines by removing the four front roller sections, installing a different axle to support regular front wheels outside the roller yoke, and attaching grouters to the drivers. Case also manufactured rock crushers, graders, grading plows, dump and spreader wagons, scrapers, and scarifiers.

This cut of a Case steamroller is unusual, for the engine has no canopy. Of greater interest is the fact that the bunkers lack the famous decal depicting the factory scene. Also note the prodigious size of the eagle transfer. Case rollers could be fitted with deep picks on the driver wheels as illustrated here. *Erb collection*

This photo depicts an early Case 10-ton steamroller (ca. 1910) climbing Center Street hill in Racine, Wisconsin, where Case products were manufactured. The smokestack on this roller is typical of early models. All Case rollers came equipped with power steering. *Erb collection*

This picture not only shows a clear shot of the Case scarifier but also depicts how effective these devices were, as may be detected in the road surface in the foreground. Of interest is how the Case scarifier and steam cylinder are offset, affording the engineer easy access to the platform. *Erb collection*

At first glance, this picture appears to have been taken around 1910, but it actually was snapped in the 1940s. It shows an early restoration of a Case steamroller. It is thought that this roller, at the Case plant in Racine, may have served as part of Case's centennial in 1942. *Erb collection*

This photograph depicts the same roller seen on the previous page after it was unloaded from a lowboy trailer. The Case rollers are modified 40-horsepower farm engines. Of note is the four-piece front roll, said to be made up from four 40-horsepower flywheels. *Erb collection*

The photo shows four products manufactured by the J. I. Case Company: (from right) a 40-horsepower Case steamroller, a water wagon, a grader, and, at the far left, a Case automobile. The conventional wisdom was that a Case farm implement dealer could sell not only road-building equipment but also the autos that would drive on those roads. *Erb collection*

This rare view of a Case roller with its curtains down shows that the Case curtains covered virtually the entire machine. As with other roller manufacturers, the brand name is stenciled on the side curtains. *Erb collection*

Not only could the center rolls be removed to convert a Case roller into a hauling engine, but the company also sold a special front yoke and wheel arrangement that accomplished the same purpose. It is noteworthy that the smokebox door in most of these original pictures is not painted black. Case priced its 10-ton roller at $2,200 with simple cylinder and at $2,300 with compound cylinder. The firm's 12-ton rollers were priced at $2,500 and $2,600 depending on which cylinder arrangement the customer ordered. *Erb collection*

The Coldwell company began in the early 1900s and manufactured steamrollers until the early 1920s. The firm later converted its steamrollers into gasoline-powered rollers. Comparisons have been made between the tandem roller owned by the Rough and Tumble Engineers Association and two gasoline-powered Coldwell machines known to exist. With the exception of the engines, the drive line and chassis are identical across all three machines. Coldwell rollers were marketed through home-improvement magazines and designed for use on large estates. They came in both a roller form and a roller/lawn mower combination. It is believed that this company ceased operations in the early 1920s.

A left-hand view of the very rare Coldwell steam lawn roller and mower combination machine. The engine is located on the front side of the boiler and is vertically mounted. This is a Mason brand of engine and is identical to those used on early Stanley Steamer automobiles. This machine, now part of the Rough and Tumble Collection of Kinzers, Pennsylvania, is said to have been purchased during the Teddy Roosevelt presidency for the lawns of the White House. *Drake Collection*

The Enright Road Roller, San Jose, California

Irishman Joseph Enright constructed his first portable threshing engine in Napa, California, in 1860. Four years later, he opened a foundry in San Jose. In 1875, 1878, and 1881, he received patents for his straw-burning boiler. By 1881, he had manufactured over two hundred portables. In 1882, Enright built a road engine with drive wheels having a sixteen-inch face. In 1885, he demonstrated "Billy the Masher," his new steamroller, to the San Jose City Council.

This picture shows the Joseph Enright steamroller known by the name of "Billy the Masher." In 1885, it was demonstrated before the San Jose City Council. From the overall size, the authors suspect that this was an extremely heavy roller. *Sourisseau Academy, San Jose State University*

This company was in business from 1886 to 1948 manufacturing woodworking products. The firm built sawmills, wood lathes, saws, surfacers, jointers, planers, drum sanders, feed grinders, carriages, boilers, and portable engines. In 1898, the Enterprise Manufacturing Company produced the three-wheel "Columbian" steamroller. How many were built and for how long remain mysteries; little information on the history of this company exists.

This right-hand view of a Columbian steamroller shows its derby-wearing engineer proudly posing for the camera. The large, solid flywheel is similar to that used on Kelly-Springfield rollers. *Drake Collection*

This 1898 illustration depicts the very rare Columbian steamroller. This machine is equipped with power (friction) steering and has a front roller yoke that was later used on Huber and Port Huron steamrollers. *Drake collection*

This photograph, made in 1898, was, according to Columbiana historians, taken in that city in Ohio. This very rare steamroller has the engineer standing alongside holding his boiler pipe wrench, indicating that he may have been making minor adjustments, as was often necessary. *Drake collection*

Erie Machine Shop, Erie, Pennsylvania

Erie native Philip W. Dietly, son of Swiss immigrants, founded the Erie Machine Shop. He began his career as an apprentice machinist then moved on as an engineer on the Anchor Line of lake steamers before starting his machine business in 1885. The company ceased operations in 1950. The firm was engaged in the manufacture of boilers and stationary steam engines, eventually building asphalt mixers and tandem steamrollers. In 1909 the company claimed to be the second American firm to manufacture rollers, but the authors have not been able to substantiate this assertion.

Shown is an 8-ton Erie steamroller compacting pavement on the shoulder of a road. The double staggered step gear can be clearly seen mounted on the inner rim of the rear drum. Such early machines are virtually carbon copies of the Lindelof roller, suggesting that they were either built under license or, more than likely, copied without any royalties having been paid to the inventor. *Historic Construction Equipment Association Library*

This street scene shows the three sizes of rollers that the Erie company marketed. At the left is an 8-ton model, in the center is a 5-ton roller, and at the right is a 10-ton machine. A 5-ton machine delivered a compression of 185 pounds per square inch, an 8-ton machine delivered 275 psi, and a 10-ton machine delivered 300 psi. While state-of-the-art in the 1890s, this type of roller was obsolete by 1910; yet, it is interesting to note that the Erie firm was able to market this design as late as 1925. *Historic Construction Equipment Association Library*

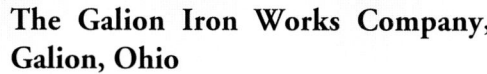

The Galion Iron Works Company, Galion, Ohio

In 1907, six entrepreneurs founded the Galion Iron Works. The firm began making iron culvert pipe. Soon, it diversified by manufacturing roadbuilding equipment, such as drags, scrapers, sprinkling wagons, stone spreaders, portable gravel screening plants, and rock crushers. Several companies advertised Galion equipment. In 1913, the name became the Galion Iron Works & Manufacturing Company. In 1916, Galion began experimenting with a gasoline roller. The first production model appeared in March 1921. The two-cylinder gasoline motors on the early Galion rollers were hard to start, so the firm began producing macadam steamrollers in 1922. The following year, Galion added heavy tandem steamrollers. In 1924, Galion reverted to gasoline, using Fordson tractor motors.

At the time of writing, Richard Rasmussen owned the only known Galion roller still in operating condition. It was serial number 202. Serial number 235 was privately owned in Pennsylvania, and serial number 284 was at the Galion Historical Society.

A late arrival to the compaction business, Galion began by producing horse-drawn graders, drags (a type of scraper), and water wagons. From about 1910 to 1917, the firm was a distributor for Baker steamrollers, and, by 1920, the company was selling Russell road rollers. It was in 1920 that Galion introduced its first steamroller. The firm started building gasoline-powered rollers in 1916, and, because such rollers were not a commercial success, the company chose to enter the steamroller business. The Galion is basically a rehash of designs pioneered by others. While most steamrollers had a maximum working steam pressure of 150 psi, the Galion used 200 psi. By 1922, Galion returned to the internal-combustion engine and abandoned steamroller manufacturing. These two pictures show good views of the Galion three-wheel steamroller. *Lantz collection*

This left-side view of a Galion 7-ton tandem steamroller is virtually a carbon copy of the firm's competition just down the road: the Buffalo-Springfield steamroller. This photo was taken in 1923 at about the same time that Galion quit the steamroller business. *Drake collection*

Here may be seen a right-side view of the same Galion tandem steamroller. Notice that the pin-striping pattern is remarkably similar to that used on Buffalo-Springfields. *Drake collection*

Geiser

Geiser Manufacturing Company, Waynesboro, Pennsylvania

Peter Geiser received a patent for a thresher, cleaner, and conveyor in 1852. He had begun work on his thresher in 1848, making Geiser one of the first to perfect a machine that separated and cleaned the grain. By 1855, Geiser was manufacturing threshers at the Geiser farm near Smithsburg, Pennsylvania, and had entered into an agreement with Jones & Miller of Hagerstown, Maryland, to build threshers. In 1858, Geiser erected threshing machines at George Frick's shop in Ringgold, Maryland. When Frick moved to Waynesboro, Pennsylvania, in 1860, Geiser set up business on part of Frick's land purchased two years earlier. Frick continued selling Geiser threshers. In the dark financial year of 1861, Geiser's friend A. B. Farquhar persuaded J. I. Case to take another block of Western territory for which Case paid Geiser the windfall of $1,100. In 1866, Geiser, Price and Company formed. In 1869, the Geiser Manufacturing Company incorporated.

In 1879, the firm bought the steam engine works of brothers Frank F. and Abe B. Landis of Lancaster. It took time to move the machinery to Waynesboro. Geiser built its first Peerless traction engine in 1881. At the Cincinnati Industrial Exposition, the Peerless took first place—quite an achievement considering that the engine pulled the water wagon, threshing machine, and maintenance vehicles from Waynesboro to Cincinnati.

The firm built threshing machines, rice threshers, clover and alfalfa hullers, hay presses, horse sweeps, gang plows, sawmills, steam engines, steam plowing equipment, and the 10-ton Peerless road roller. In 1912, the Emerson-Brantingham Company of Rockford, Illinois, bought the Geiser company. E-B sold Geiser to Waynesboro investors in 1925. The stock market crash of 1929 crippled the company. The firm went bankrupt in 1939.

THE GEISER MANUFACTURING COMPANY INC.

PEERLESS 1910

WAYNESBORO, PENNA., U.S.A.

This 1910 catalog cover shows a front view of a Geiser "Peerless" road roller. The canopy is fitted with a canvas flap like those found on awnings. This is another rare brand of steamroller. *Rhode collection*

A right-hand shot of the Peerless steam-roller. Unlike the roller depicted in the catalog cover, this machine is fitted with side curtains, not an awning flap. The steering mechanism with the angled smokebox door is unusual in American practice. *Geiser catalog*

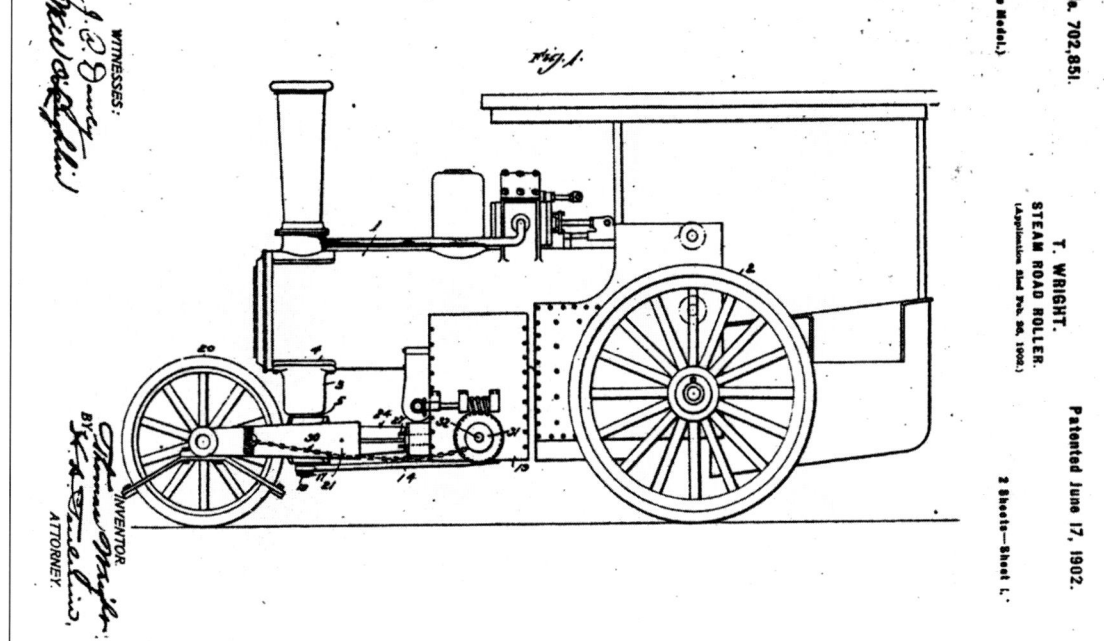

While this 1902 drawing accompanied a patent by Thomas Wright, it is included here because of its similarity to the Peerless design. To the authors' best knowledge, no other brand of roller followed this pattern.

The Harrisburg Car Manufacturing Company, Harrisburg, Pennsylvania

In 1853, eight citizens of Harrisburg, Pennsylvania, and a railroad car builder from Worcester, Massachusetts, founded the Harrisburg Car Manufacturing Company. The Civil War's demand for rolling stock stimulated production. In 1866, the car works and machine works expanded and became the Harrisburg Foundry and Machine Works. The Panic of 1873 menaced the future of both businesses. By 1875, the car company had righted itself financially. It took over the unsteady foundry's assets and closed that plant. In 1879, the rosy economic picture encouraged Harrisburg Car to reopen the foundry. The plant not only machined wheels and parts for railroad cars but also made oil tanks, gas flues, standpipes, heavy equipment for rolling mills, machinery for blast furnaces, boilers, compound pumping engines, traction and portable engines, farm implements, sawmills, and steamrollers. The factory annually turned out 150 of the Paxton Portable Steam Engines, named for the Little Paxton Creek that cut across the plant's property.

In 1887, Harrisburg Car began manufacturing steamrollers, and, at the same time, the firm's farm engine business began to decline. According to Gail E. Anderson of STEMGAS Publishing, "Public confidence had been shattered by scandals uncovered in the banking and railroad securities exchange professions. Railroad companies were in no position to invest in new material or rolling stock, which left the Harrisburg Car Manufacturing Company with very little to do." In 1890, a bank failure led to the dissolution of the Harrisburg firm. Harrisburg Car's three-wheel steamrollers had been the first modern three-wheel rollers produced in America,

with the word "modern" differentiating the design from that of Abbott Q. Ross. After the firm ceased operations, Martin E. Hershey, the inventor of the Harrisburg Road Roller, filed for a three-wheel roller patent on February 18, 1891, and received it on July 28, 1891. Hershey barely beat Edward T. Wright, the inventor of O. S. Kelly's three-wheel steamroller. Wright filed for his patent on March 23, 1891, and received it on December 29, 1891. Harrisburg had gone out of business, but the authors discovered that the O. S. Kelly Company, from June 16, 1904, to February 7, 1905, built seven Harrisburg 10- and 12-ton steamrollers. This fact is a mystery wrapped in an enigma.

This early cut from *Hawkins's Catechism* shows a detailed shot of the rear of a Harrisburg steamroller. The smokestack was taller and was cropped off by the printer. Harrisburg rollers are very rare. *N. Hawkins's New Catechism of the Steam Engine (1904)*

Fig. 248. STEAM ROAD ROLLER.

New Catechism of the Steam Engine. — 353

This scene, taken in the Queen Anne suburb of Seattle, Washington, shows a Harrisburg roller working next to interurban streetcars. This roller is fitted with a half cab canopy identical to that used on the 1892 vintage O. S. Kelly steamrollers. Of note are the steel picks installed in the rear wheels for additional traction in the muddy streets. *Rhode collection*

Taken in Massachusetts during the 1890s, this picture shows a large Harrisburg roller. Unlike most others, this roller is not equipped with a belly tank for extra water storage. In addition, a very noticeable weak point is the front wheel steering yoke that consists of riveted steel. Many early rollers used this type of yoke before adopting ones that were steel forgings. The riveted yokes bent easily, as has this one. *Drake collection*

Heilman

Heilman Machine Works, Evansville, Indiana

William Heilman was born in 1824 in Germany. At age nineteen, he came to the United States. In 1847, William and brother-in-law Christian Kratz built a foundry and machine shop in Evansville, Indiana. They began by making stoves and hollowware. Heilman and Kratz were fortunate. In the spring of 1848, a wave of German immigrants to Indiana brought a sudden demand for stoves. Heilman gained the capital necessary to become an industrial giant. In 1850, Heilman and Kratz built a finishing shop to manufacture steam engines and mill machinery. At first, they made sawmills. In 1852, they erected boilers in an expanded finishing shop. In 1854, Heilman and Kratz's City Foundry and Machine Works built its first portable steam engine. Before long, the firm was building steamboat machinery and large engines for Southern cotton mills. In 1859, the firm quit building stoves and began making threshing machines modeled on the Pitts brothers' patent.

During the Civil War, Heilman and Kratz sold copper stills to Southern territories controlled by Union troops. In 1864, Kratz went his own way, leaving Heilman in sole charge of the business. By 1866, Heilman had greatly expanded the plant. In 1878, Heilman was elected to the United States Congress. Through the 1880s, Heilman invested in railroads and made large profits. He died in 1890.

This illustration shows a left rear shot of a Heilman road roller. This rare machine clearly shows its farm traction engine origins. The rear wheels and bunkers are more typical of an agricultural engine than of a steamroller. T. H. Smith's *The Album of American Steam Traction Engines* (1953)

A leader traction engine, built by the Marion Manufacturing Company, helps to power a coal-fired asphalt plant produced by Hetherinton & Berner. Both Marion and H&B offered steamrollers, but the authors were unable to locate a photograph of the H&B roller. *Gencor Industries collection*

Hetherington and Berner manufactured this portable asphalt paving plant. Even with such mechanization, horses were still useful. Trinidad asphalt came in barrels, which had to be broken apart. At the right may be seen a pile of cast off barrel staves. *Gencor Industries collection*

Hetherington & Berner

Hetherington & Berner, Indianapolis, Indiana

In the 1850s, Frederick Berner came from Alsace-Lorraine to Indianapolis. In 1867, he entered into partnership with B. F. Hetherington in a small machine shop beside Pogue's Run. The firm repaired boilers and machinery. In the 1891 Indianapolis directory, Hetherington & Berner were billed as "leading manufacturers of steam boilers." The company became the first to build asphalt batch mixing plants. In 1893, the firm was incorporated. In 1897, Frederick A. Hetherington patented a portable paving plant. In 1901, Frederick Berner died, and his son, Frederick Berner, Jr., succeeded him in the business. In 1912, Frederick Berner, Jr., passed away, and his son, Robert Berner, succeeded him. Frederick and Carl Hetherington held many of the company's patents. In 1929, the Berner interests acquired the Hetherington stock. By the mid-twentieth century, the firm diversified, building batch mix sectional plants, portable asphalt plants, industrial and commercial buildings, and pumping/dredging equipment.

According to Lyle Hoffmaster, an authority on the Reeves Company of Columbus, Indiana, the three-wheel Hetherington & Berner steamroller was designed by Albert Clay, son of the brilliant mechanical engineer Harry Clay, who did much to establish Reeves's reputation for excellence.

The Huber Manufacturing Company, Marion, Ohio

In 1865, Edward Huber moved to Marion, Ohio, where he manufactured his patented revolving hay rake. He worked in space rented from the Kanable Brothers Planing Mill and borrowed many of his tools. In 1866, he served as superintendent of the newly formed Kowalke, Hammerle, Monday and Huber firm, which eventually acquired the Kanable planing mill. The Huber & Gunn Company followed in 1870. Four years later, the Huber Manufacturing Company was founded, acquiring the Holmes & Seffner machine shop. Also in 1874, Henry Barnhart approached Huber with an idea for a steam shovel. Huber backed the project, and the first steam shovel was built in the Huber shops. Later, the Marion Steam Shovel Company was founded. Huber was the president, Barnhart the manager and superintendent. Soon, Huber added a steam engine and thresher to the product list. Huber built threshers not only for wheat but also for beans, peas, and rice. The firm manufactured plows, skid engines, and portable engines. The first steamrollers were built in 1908. Huber rollers had either one or two cylinders. Like their traction engine counterparts, the steamrollers used a return-flue boiler.

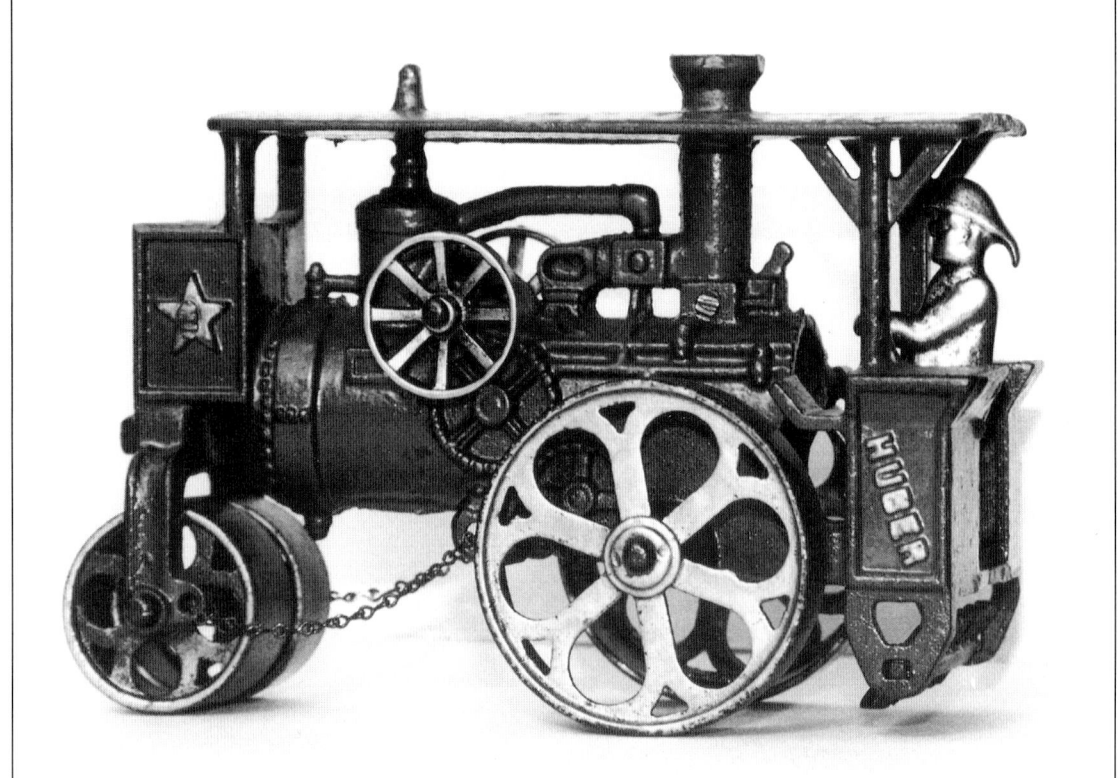

Toy models of construction equipment have always been popular with children, and steamrollers are certainly no exception to this rule. Here we see a rare and highly collectible cast-iron toy Huber steamroller, manufactured by Hubley. *Rhode collection; photo by Thomas R. Haid*

When the Huber Company chose to build a steamroller in 1908, it, like so many other builders of farm traction engines, simply modified its existing product by adding a front roll. This picture shows one of the first models marketed by Huber. *Marion County (Ohio) Historical Society*

This left rear view of the early style of Huber steamroller shows the front yoke gooseneck in great detail. The pin-stripe scheme is different than that seen in the previous picture. It also appears that the engineer's seat is mounted on coil springs. *Drake collection*

This 1923 illustration shows the right side of a Huber steamroller. This machine is the company's single-cylinder model. Unlike the earlier models that had a canvas flap around the edge of the canopy, these later rollers are fitted with side curtains. *Marion County (Ohio) Historical Society*

A left-hand view of the Huber single-cylinder steamroller. This picture shows that Huber had modified the front steering roll. This new design is similar to that used on the 1898 Columbian and the Port Huron rollers. The Huber Company claimed that this modification shortened the roller's wheel-base by nine inches. *Marion County (Ohio) Historical Society*

In this picture can be seen the left-hand view of Huber's "Top of the Line" double-cylinder roller. The Huber steamrollers were all painted battleship gray and were striped and lettered in brown. *Marion County (Ohio) Historical Society*

A right-hand view of the Huber double-cylinder steamroller. Like the company's traction engines, the road rollers used the return-flue boiler. The rollers were equipped with power steering, which can be seen in this picture. Of note is the fact that the scarifier is tucked under the rear platform. *Marion County (Ohio) Historical Society*

The Iroquois Iron Works, Buffalo, New York

This company was closely tied to the fortunes of "The Asphalt King of America," Amzi Lorenzo Barber. Born at Saxton's River, Vermont, in 1843, Barber played a dominant role in American road building. A pioneer in the auto industry, he owned the Locomobile Car Company of Bridgeport, Connecticut. His first fortune, however, came from selling District of Columbia real estate. Working in the nation's capital in 1876, he happened to witness the first ever paving with asphalt—on Pennsylvania Avenue. Barber observed that this new surfacing medium flexed and did not crack. He foresaw that asphalt would make him rich. The world's largest source of asphalt was Pitch Lake on the island of Trinidad. By 1888, Barber acquired the rights to Pitch Lake, which effectively gave him a monopoly on the world's supply of asphalt. As new deposits were found, such as those at Bermudez, Venezuela, he immediately cornered the rights to them. In 1883, he formed the Barber Asphalt Paving Company and, after purchasing the Pitch Lake properties, moved his company headquarters from Washington to New York City.

Barber decided to establish a separate company to manufacture new paving equipment. About 1890, Barber acquired the Pioneer Iron Works of Brooklyn, manufacturers of Lindelof rollers. In 1894, the roller business was moved to Buffalo and was renamed the Iroquois Iron Works. Even though the firm was engaged in the building of rollers since 1890, the company did not incorporate until March 26, 1898. By 1900, the firm added a three-wheel roller to its line. The company also built portable asphalt plants.

Barber became immensely wealthy. Besides being prominent in New York social circles, he was a well-known yachtsman. In 1909, after inspecting new pitch deposits discovered in California, Barber came down with a fatal case of pneumonia. The Barber Asphalt Paving Company survived and maintained an Iroquois department, from which, in 1920, the firm offered a startlingly new three-wheel roller.

In the *Good Roads* magazine for February 4, 1920, the Iroquois department of the Barber Asphalt Paving Company advertised this new design of a three-wheel roller having a vertical boiler.

IROQUOIS

A front view of an Iroquois three-wheel steamroller demonstrating the flexibility of the rocking joint. Pin striping and lettering are especially clear in this picture. *Drake collection*

A left front shot of an Iroquois roller fitted with a headlamp. These lamps were usually the Dietz "Champion" brand. The small flywheel and the short stroke rods in comparison to the size of the boiler suggest that Iroquois may have been underpowered. *Drake collection*

Iroquois roller right front view. These machines appear to have identification plaques on both sides of the kingpin housing. Of note are the massive kingpin housing and steering yoke used on these machines. *Drake collection*

This left-side view of the Iroquois roller clearly shows the gearing arrangement. It is interesting to note the lack of protective covering on the gears, which suggests that a person would not want to get a coat sleeve too close. Unlike other makes, these rollers had only a single speed. *Drake collection*

This picture is unusual in that rollers are seldom depicted with their curtains down. To the right of the rear wheel is the Iroquois Native American trademark. One possible design flaw is readily noticeable in that the front roll appears to be precariously close to the belly tank, seriously reducing the turning radius. *Drake collection*

This late model (1909) Iroquois roller differs from its predecessors, in that it is fitted with protective gear shields and has an unusual box arrangement on top of the engine. *Shanklin collection*

Here is a small size Iroquois tandem steamroller. These were designed to roll driveways, tennis courts, and lawns. This 2 ½-ton version uses an oiling attachment when rolling asphalt. *Drake collection*

This picture shows the left side of the Iroquois tandem steamroller. These rollers closely resemble the Lindelof products. The front yoke, this one a mixture of cast and sheet steel, is often the only way to identify the manufacturer of a roller depicted in a photograph. *Drake collection*

Iroquois tandem roller seen from the rear. These machines came with a folding metal cover over the coal tray—a nice touch especially for roller engineers who had to work in inclement weather. Iroquois offered these rollers in 2 1/2-, 3-, 5-, 7 1/2-, 8-, and 10-ton sizes. *Drake collection*

IROQUOIS

As a division of the Barber Asphalt Paving Company, Iroquois also made Barber's asphalt batching plants. In this picture is their railroad mounted portable asphalt plant. *Drake collection*

Like most other roller manufacturers, Iroquois also built additional products related to the paving industry. Here is the firm's steam-powered concrete mixer, the boiler and engine of which appear to have been borrowed from the company's tandem rollers. *Drake collection*

This picture shows the semi-portable asphalt plant that was often purchased from the Iroquois Company by municipalities and smaller contractors. *Drake collection*

Kelly

Oliver Smith Kelly struck it rich in the California gold fields between 1852 and 1856. He returned to Ohio and invested in a wholesale grocery in his hometown of Springfield. In 1857, Kelly joined the farm implement firm of William Whitely and Jerome Fassler. The name soon changed to Whitely, Fassler & Kelly. In 1882, Kelly invested in the Rhinehart and Ballard Threshing Machine Works, which sold the John Pitts threshers. Shortly after (in 1882) the firm reorganized as the Springfield Engine and Thresher Company. About 1890, the name changed again when the business incorporated as the O. S. Kelly Company. Edward T. Wright, a British subject in Springfield, taught Kelly the advantages of British steamers. Kelly's designers began to replace the Springfield engine with a new style of engine modeled on British concepts. Kelly opened a branch in Iowa City, Iowa. He appointed James H. Maggard General Manager. Steam aficionados prize their copies of Maggard's popular book, *Rough and Tumble Engineering*. From 1898 until approximately 1905, Kelly experimented with a triple-cylinder, cross-compound, 120-horsepower, cable-plowing engine. Some were shipped to Cuba as road locomotives.

In 1892, Kelly produced what is often mistakenly said to be the first self-propelled steamroller in the United States. Founded in 1902, the Kelly-Springfield Road Roller Company, an outgrowth of the O. S. Kelly Company, later merged with the Buffalo Steam Roller Company to form the Buffalo-Springfield Roller Company. Oliver's son Edwin joined inventor Arthur W. Grant in founding the Rubber Tire Company, forerunner of the Kelly-Springfield Tire Company. Early in the century, Kelly experimented with a steam truck. In 1910, Edwin organized the Kelly-Springfield Truck Company. Kelly also built plates for world-class grand pianos.

This 1892 engraving is the earliest known image depicting an O. S. Kelly steamroller. It appears to be one of the larger rollers manufactured by Kelly. This is the only picture of a Kelly to be equipped with a half-cab canopy. This roller has not only the square headlamp but also a similar lamp located on the rear platform, probably used as a taillight. *Drake collection*

This is the earliest known advertisement for a Kelly roller and is from the May 1893 publication *The Review of Reviews*. *Rhode collection*

This 1893 picture of a large Kelly steam-roller shows the power steering unit invented by the firm for use on rollers weighing over fifteen tons. This photograph shows the paint stripe pattern employed not only on such early rollers but also on Kelly farm engines. These early machines had single-cylinder engines and one-speed transmissions. *Drake collection*

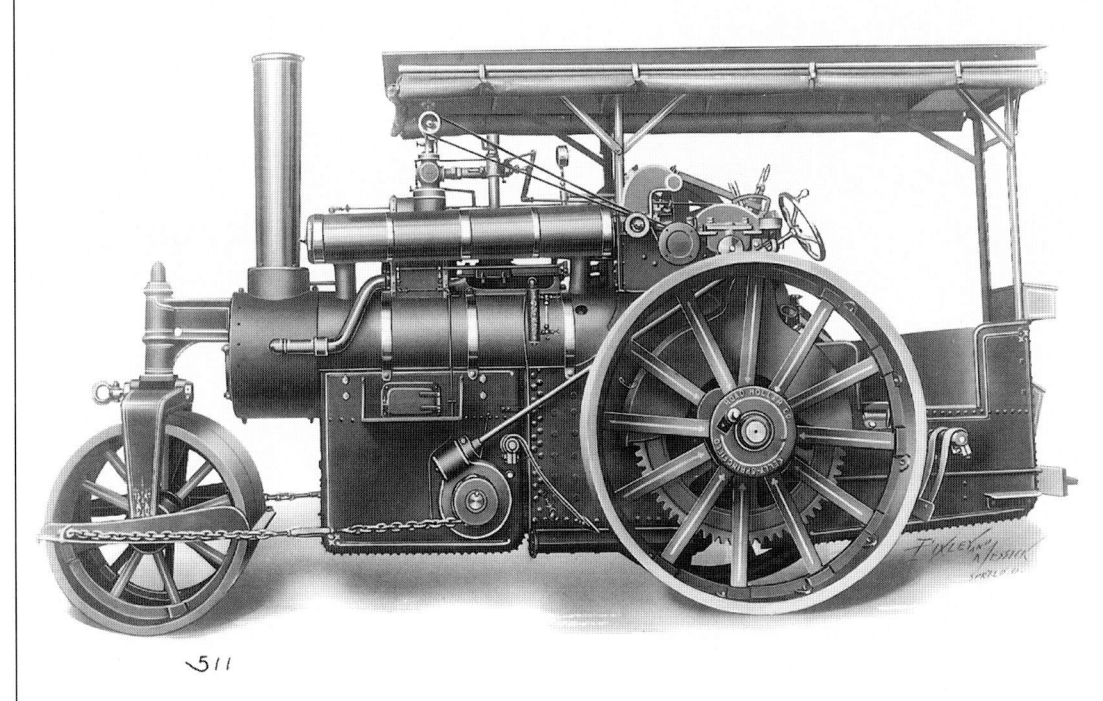

This 10-ton Kelly shows the paint and striping combination commonly employed (with minor variations) from 1906 until the end of production. The roller was painted Brewster green, and the striping was gold. This roller has the horizontal steam chamber that—although a hallmark of such rollers—was not used on the very early models. After 1906, Kellys also came with double-cylinder engines and two-speed transmissions. *Lawrence Clark collection*

Kelly-Springfield model SM33 "Fortified," a 10-ton machine. The plaque just below the smokestack indicates that this roller was made for the Dutch exporter Lindeteves-Stokvis. This firm had its main offices in New York City and Amsterdam, Holland. The company also had branch offices in the Dutch East Indies (present-day Indonesia) in the cities of Semarang, Surabaya, Batavia, Tegal, Jakarta, Bandung, Medan, and Makassar. With all these branch offices, one wonders how many steamrollers survive in Indonesia today. *Drake collection*

This picture shows Kelly-Springfield, serial number 4836, a 12-ton roller built on June 26, 1920. There are small differences between this machine and those that were built before the merger with the Buffalo Steam Roller Company in 1916. There is also the major difference that this roller is equipped with a scarifier. This machine came with 20" rear wheels, which were standard on the 12-ton models. The 10-ton models had 18" wheels, and the 15-ton types had either 22" or 24" wheels. *Crout collection*

Taken August 4, 1926, this picture shows a later model 10-ton Kelly-Springfield that was designed for reservoir or steep embankment work. Kelly roller serial number 1951, a 5-ton tandem manufactured on July 28, 1908, was the first to be fitted with reservoir type rolls. *Lavender collection*

In this picture, the basic components of the Kelly steamroller can be seen: the boiler with the gears and shafting. This is a post-1910 photograph, in that this engine has a Pickering governor, whereas the early machines used the Waters governor. The Kelly rollers employed a Marshall valve gear, while the Buffalos had Stephenson valve gearing. *Drake collection*

Kelly rollers were used on many famous construction projects, including the Panama Canal. In this 1911 picture may be seen a Kelly embankment or reservoir roller building the Indianapolis Motor Speedway. *Drake collection*

Here is a left front view of the Kelly reservoir, or puddling, roller. This August 1926 machine has a striping design that is almost identical to that of its cousin, the Buffalo-Springfield. It no longer has a Kelly trademark on its belly tank; instead, it now says "Buffalo-Springfield" and has been relegated to being just another model for that firm. This change appears to have taken place in 1925. *Lavender collection*

This photograph, taken in 1918 during the construction of a reservoir for the New York Water Works, shows how a steamroller was brought into, and removed from, a construction project that had steep embankments. Here, a large Kelly roller, probably a 16-ton model equipped with power steering, is lowered into the reservoir. The roller was transported by railroad to the derricks in the distance, where it was pulled off the flatcar by cables connected to steam donkey engines located behind the derricks. Using heavy steel cables that ran over a sheave wheel at the top of the derrick and across the work area to derricks on the opposite bank, the crew could pull the roller to the desired location, then lower it to the floor of the reservoir. When the project was completed, the roller was removed in the same manner. *Clark County (Ohio) Historical Society*

This picture was also taken in 1918 during the construction of the New York Water Works. This machine is the 20-ton model and is equipped with a large tank located on the rear platform, indicating that this roller is fired with a liquid fuel. The large-size Kellys can be identified by the two-piece smokestack and also by the fact that they have two safety valves, as evidenced by the two pipes that come up through the front of the canopy. *Drake collection*

The last Kelly-Springfield three-wheel steamroller to be built. This machine, serial number 13415, a 12-ton roller, was from job #263, started on November 16, 1929, just after the great stock market crash, which led to a drastic drop in sales of all types of construction equipment. As a result, this roller was not finished until March 17, 1934. With the election of Franklin Delano Roosevelt in 1932 came New Deal public work programs that required construction equipment, including rollers, which gave companies like Buffalo-Springfield new markets. *Clark County (Ohio) Historical Society*

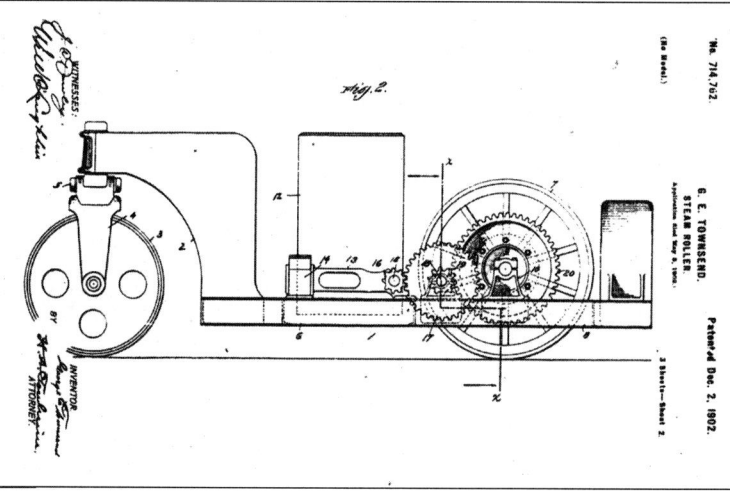

The Kelly Company realized that, if it were going to sell tandem rollers in large numbers, it would have to solve the rocking problems associated with that type of roller. This page traces the evolution of the Kelly tandem roller.

(top left) . . . 1891 woodcut of the first Kelly tandem steamroller.

(top right) . . . 1893 model with the boiler moved to the side allowing the engine to be slanted, which reduced but did not eliminate the rocking motion.

(bottom left) . . . first production 4-ton tandem roller. Built in 1904, it came with the new engine and reduction gear design. Also, this roller was the first to have roller bearings in the hubs of its rolls. *Drake and Rhode collections*

(bottom right) . . . December 2, 1902, patent drawing for the Kelly horizontal engine and reduction gearing design.

Kelly-Springfield and Buffalo-Springfield both offered four basic sizes of tandem steamroller. After 1916, without checking the serial number, it is virtually impossible to tell the difference between a Buffalo or Kelly roller. In this 1917 picture may be seen (from left to right) a 10-ton, a 7-ton, a 5-ton, and a 2 ½-ton steamroller. *Clark County (Ohio) Historical Society*

This is the standard design for Kelly tandem rollers after 1908; it remained so throughout the end of production. While there were four basic sizes, the company varied those sizes with the addition of heavier wheels. For example, a 5-ton could be turned into a 6-ton by substituting heavier wheels. *Drake collection*

Kelly medium-sized rollers were popular with contractors who built brick roads. This roller is unusual, in that it is fitted with a belt pulley. While belt pulleys were common on three-wheel rollers, this is the only picture of a tandem roller fitted with a belt pulley that the authors have encountered. *Drake collection*

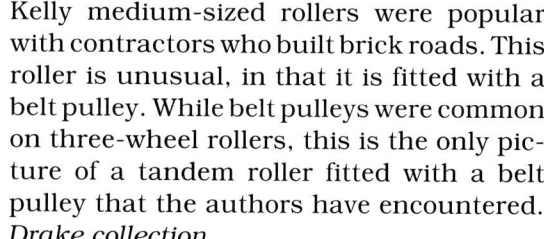

At the beginning of the twentieth century, there were literally hundreds of people trying to break into the automobile business, and, to a lesser extent, this was also true in truck manufacturing. In 1903, the O. S. Kelly Company built this steam-powered truck, seen in May of that year hauling 11,000 pounds of cast-iron pipe on the streets of Springfield, Ohio. *Rhode collection*

The Kelly truck had its steam engine mounted on the rear axle, along with having its water storage tank behind the rear axle. The boiler was placed at the center of the chassis and was fired with liquid fuel from the storage tank located under the driver's seat. These steam trucks were not a commercial success; thus, when Kelly in 1910 reentered the truck business as Kelly-Springfield, all the firm's trucks were powered by internal-combustion gasoline engines and would remain so until the late 1920s, when the company ceased production. *Rhode collection*

Anders Lindelof, Brooklyn, New York

On February 11, 1873, Anders Lindelof received a patent for the first American tandem gooseneck steamroller. Lindelof had begun manufacturing his roller in 1871—the same year that Abbott Q. Ross of Cincinnati patented his machine—but Lindelof's patent came later. Born in Sweden in 1837, Lindelof immigrated to America in 1865 and remained in Brooklyn until his death in 1915. While he patented and built some of the first tandem steamrollers to be made in America, these early machines had problems. First, the large two-cylinder engine was directly connected to a drive shaft and pinion gear that drove a large ring gear around the inside of the rear drum. This direct drive feature meant that the crankshaft attained only about seventy revolutions per minute. The long piston rods caused a rocking action, leaving a wavy pattern on the rolled surface. In 1902, the Kelly-Springfield Company corrected this situation by patenting the horizontal engine. Another problem was that the front roll had a diameter of twenty-six inches, too small for a roller of this size. The roll tended to dig into the asphalt and push it ahead, leaving a depression. A roller of this magnitude (5- to 6-ton) needed a roll with a minimum diameter of thirty inches. Despite these early problems, the tandem roller is today the standard of the industry.

Patented in February of 1873 and first shown at the Centennial Exposition in 1876 held in Philadelphia, this was one of the first tandem rollers to be built and marketed in America. Lindelof rollers can be distinguished by their pressed steel front yoke and small front roll—only twenty-six inches in diameter. These were produced in small numbers until 1890, when Lindelof sold his patent rights. *Vogel collection*

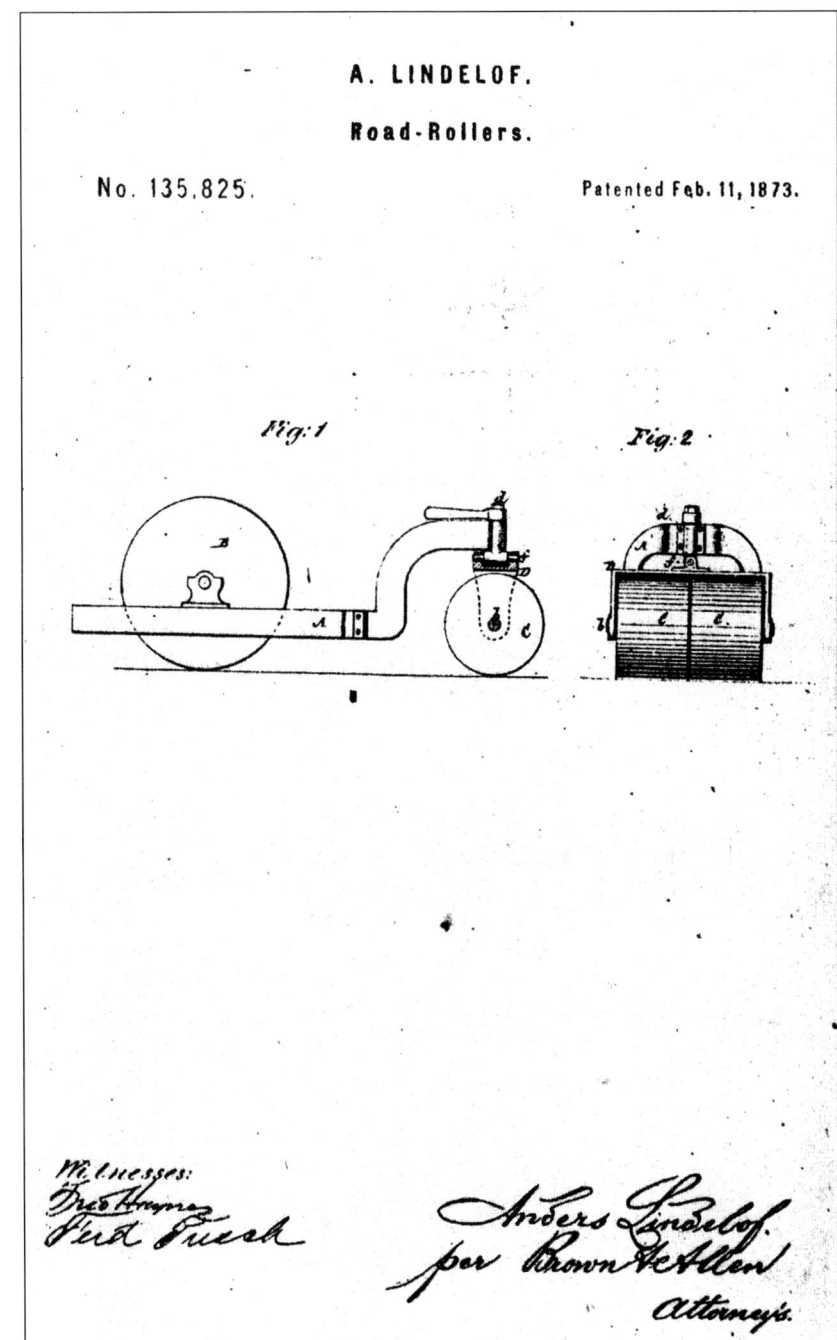

A. LINDELOF.

Road-Rollers.

No. 135,825.

Patented Feb. 11, 1873.

Fig. 1   Fig. 2

Witnesses:

Anders Lindelof
per Brown & Allen
Attorneys.

For over a hundred years, Anders Lindelof has been called the father of the American steamroller. During the research for this book, the authors discovered that Abbott Q. Ross had taken out a patent a few years before Lindelof. Both built steamrollers during the 1870s, but who built the first machine remains a mystery. Although Lindelof was the second to file his patent, his contribution is not to be minimized because it was his design that was adopted by the compaction industry. *Original patent drawing courtesy of John C. Lindelof*

Marion

The Marion Manufacturing Company, The Leader Company, and The Ohio Tractor Company, Marion, Ohio

In 1886, the Marion Manufacturing Company formed. According to rumor, a squabble at the nearby Huber Manufacturing Company caused disaffected Huber employees to found the firm. The manufacturer made sawmills; threshers; and skid, portable, and traction engines. Late in its existence, the company also produced a 10-ton steamroller. The farm engines built by Marion were known by the trade name of "Leader." Toward the end of the company's history, the firm became known as "The Ohio Tractor Company," and all three names—Marion, Leader, and Ohio—appeared on company stationery. According to T. C. Spires, who owned Leader engines, Warren Gamaliel Harding, the twenty-ninth President of the United States, once had an interest in the Ohio Tractor Company, perhaps presiding over it. The firm went bankrupt about 1917.

Here again may be seen an instance of a manufacturer adapting its farm traction engine to become a road roller. This firm has fitted its product with proper roller wheels, and the short wheel base is an added advantage. Like the Huber and Port Huron rollers, this machine has a front steering yoke similar to that introduced on the Columbian roller in 1898. *Reynolds Museum*

In this ca. 1914 photograph may be seen the principal buildings of the Monarch road roller plant at Groton, New York. As with all manufacturers of heavy construction equipment, shipping of the finished product was of vital importance. Here are railcars of the Lehigh Valley Railroad waiting to be loaded with new Monarch steamrollers. At the time of this writing, a retirement village occupies this site. *Wittmer collection*

The Groton Manufacturing Company, The Conger Manufacturing Company, The Monarch Road Roller Company, and The American Road Machinery Company, Groton, New York

In 1849, brothers Charles and Lyman Perrigo established a foundry in Groton, New York, to manufacture agricultural machinery. Approximately a decade later, William Perrigo joined S. Spencer to produce threshing machines. In 1863, William Perrigo bought out Spencer. William Perrigo and Frederick Avery joined Charles Perrigo and Company. In 1887, the firm became known as the Groton Bridge and Manufacturing Company, building iron bridges, portable steam engines, grain separators, and steam and hot air heaters. The name of Charles Perrigo appeared on many of the firm's products, including at least one traction engine. By the late 1890s, the company was manufacturing Groton engines and Monarch road rollers, and no products bore the name of Charles Perrigo. It is unclear if any road rollers appeared under the Perrigo name. From 1899 through 1901, the bridge building segment of the company was spun off under separate management, called the American Bridge Company. The traction engine and road roller portion of the business became the Groton Manufacturing Company. In late 1901, the firm changed to the Conger Manufacturing Company. Both the earlier Groton Manufacturing Company and Conger built traction engines and road rollers under the Groton and Monarch names. During August of 1903, the firm was known as the American Road Roller Company and also as the Wright Roller Company. At this time, William C. Oastler, long the American distributor of British Aveling & Porter rollers, came into control of the company, and, simultaneously, a new model of steamroller was introduced that was quite similar to the Aveling & Porter design. From 1907 to 1910, Charles Longenecker & Company of New York offered a Monarch three-wheel steamroller featuring an improved bunker design based on Longenecker and Edward T. Wright's patent. In September of 1903, the Monarch Road Roller Company emerged, retaining that name until 1914, when it was absorbed by the Good Roads Machinery Company, calling its manufacturing branch the American Road Machinery Company. The Monarch name appeared on the firm's rollers until the Groton plant closed in 1925. While Groton served as the company's manufacturing center, the American Road Machinery Company corporate headquarters were at Kennett Square, Pennsylvania, near New Jersey and that state's powerful movement for good roads. The words "Good Roads" were cast into the valve-chest cover on Monarch tandem rollers. The Monarch three-wheel steamroller could be converted into a road locomotive with two front wheels.

A rare interior view of the Monarch machine shop. The machine on the right is a large vertical lathe that is machining the face on a rear wheel. *Wittmer collection*

This picture shows the final assembly area at the Groton plant. The unusually large cast kingpin and smokestack housing suggests that these are early model rollers. It is interesting to note just how clean and tidy the work areas appear in both the machine and assembly shops. *Wittmer collection*

These two pictures reveal the interiors of the Monarch foundry and boiler shops. On the right of the foundry picture can be seen freshly cast engine frames. In the boiler shop are square objects on the right that are boiler mud rings. Behind the rings are boiler tubes. *Wittmer collection*

This picture shows a left-hand view of a Monarch "King of the Road" steamroller. This machine has its rear wheel set on top of a twenty-inch piece of oak to show the flexibility of the chassis. This shot also gives a good view of the axle gears. Monarch rollers were unique, in that they came with differential gears. *Wittmer collection*

Driving Wheel
(*Left Side of Machine*)

Driving Gears
(*Right Side of Machine*)

This illustration, taken from an old Monarch catalog, clearly shows the differential mechanism complete with spider gears. *Wittmer collection*

These two illustrations depict the types of scarifiers used on Monarch rollers. The first, or early design, is manually operated and folds under the rear water tank when not in use. The second, or later type, is similar to the Buffalo-Springfield design. This must have been rather impractical, as the Monarch is a rear entry machine, which would force the engineer to climb over the scarifier cylinder. *Wittmer and Owensby collections*

This picture shows the right rear shot of a "King of the Road" roller. This machine is equipped with a belt pulley, often used to operate rock crushers. Note the unusual support bracket that is attached to the axle shaft. *Wittmer collection*

476

A left-hand view of Monarch tandem roller. This type of roller, with its upright engine, had several drawbacks, not the least of which was its direct drive ring and pinion gearing. The upright engine with its direct drive gearing caused a rocking action that left a wavy pattern on the pavement—a very undesirable situation. While Kelly-Springfield patented the horizontal engine with reduction gears, thereby eliminating the wavy pattern, some companies continued to make obsolete designs well into the 1920s. *Wittmer collection*

Monarch also built tandem rollers in 6-, 7-, 8-, and 9-ton sizes. This machine, as well as the tandem machines built by Iroquois and Erie, is a virtual carbon copy of the Lindelof roller. With the exception of nameplates, the only easy way to distinguish the different brands is by closely examining the front steering yoke. On Monarch rollers, they are all cast. *Wittmer collection*

This is the Monarch "Combination" road roller, which was an adaptation of the firm's farm traction engine. The major difference between this roller and the "King of the Road" model can be seen at the front end. While the "King of the Road" has a large kingpin housing, similar to that of a Buffalo-Springfield, this machine employs a large thrust washer steering fork, very similar to that used on Case rollers. *Wittmer collection*

This right-hand view of the "Combination" roller clearly shows just how it could be converted into a road locomotive or hauling engine. Traction lugs were placed on the rear wheels, the two center rolls were removed, and a spacer was placed on the axle, after which anti-skid rings were bolted to the front wheels. The result was a traction engine for haulage purposes. *Wittmer collection*

A left rear view of the Monarch "Combination" roller shows a detailed shot of the engine. Also notice that this machine has a steering mechanism that is different from the "King of the Road" models. *Wittmer collection*

An interesting view that shows the heart of all steam rollers—the boiler. The axle shaft and main gear shafts have been installed. These shafts were all set in phosphor-bronze bearings. *Wittmer collection*

Port Huron

The Port Huron Engine and Thresher Company, Port Huron, Michigan

In 1882, the Upton Manufacturing Company sold its first traction engine. In 1884, the factory moved to Port Huron. In 1890, the company's name changed to the Port Huron Engine & Thresher Company. The following year, the Port Huron Rusher Thresher and Port Huron engine appeared. A few of the first "Longfellows"—engines with boiler tubes nine feet long—were introduced in 1907. By 1908, Longfellows had proven themselves. Port Huron built sawmills; hay presses; special wagons for hauling; corn shellers; threshers; water tanks; skid and portable engines; single-cylinder, tandem-compound, and double-tandem-compound traction engines; large steamrollers that could serve as hauling engines; regular road rollers, and the "New York Huron Standard" steamrollers.

The Port Huron steamroller was simply a 24-horsepower farm traction engine conversion. As with the Marion and Huber rollers, this machine employed the center mount steering fork first used by the Columbian roller in 1898. *Port Huron catalog*

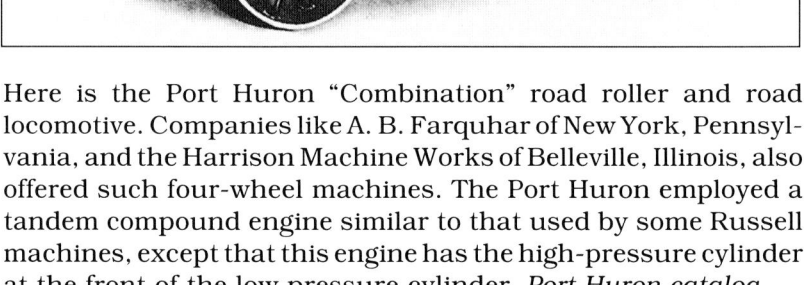

Here is the Port Huron "Combination" road roller and road locomotive. Companies like A. B. Farquhar of New York, Pennsylvania, and the Harrison Machine Works of Belleville, Illinois, also offered such four-wheel machines. The Port Huron employed a tandem compound engine similar to that used by some Russell machines, except that this engine has the high-pressure cylinder at the front of the low pressure cylinder. *Port Huron catalog*

This 1890 illustration of the Jacob Price Field Locomotive has never been published before. According to the catalog of that year, the Field Locomotive could be modified into a road roller with the addition of smooth wheels. The dark areas on this rare illustration are the result of repairs done with plastic tape. *The F. Hal Higgins collection, Department of Special Collections, University of California Library at Davis*

The Jacob Price Road Roller, Racine, Wisconsin, and San Leandro, California

In 1887, Jacob Price of San Leandro, California, patented a traction engine. He built a steamroller designed on his patent and demonstrated it in San Jose, California, in the same year. The roller performed poorly, and the San Jose *Mercury* minced no words in saying so. Explaining that the roller had not been tested before being shipped to San Jose, the undaunted Price in 1889 turned to the J. I. Case Threshing Machine Company to build a field locomotive/steam plow based on his latest ideas. He entered the Racine-built engine in the California State Fair plowing contest in 1890, the same year that he received a patent for his engine. By a narrow vote, the judges chose Daniel Best's engine over Price's. Case made several engines for Price. In 1893, Price signed over his patents to Case to settle his debts with the Racine firm. The following year, Case and the related company called the J. I. Case Plow Works issued a catalog to promote sales of the engine. By 1897, Case and the Plow Works had ceased production of the Price engine.

Julian Scholl

Julian Scholl and Company, New York, New York

Julian Scholl acquired the rights to the Charles Heisler and W. M. Fawcett patented tandem roller, which he manufactured in limited numbers in the early years of the twentieth century. His unusual machines were called "The Universal Roller." Scholl was the New York City distributor for Russell and Iroquois steamrollers.

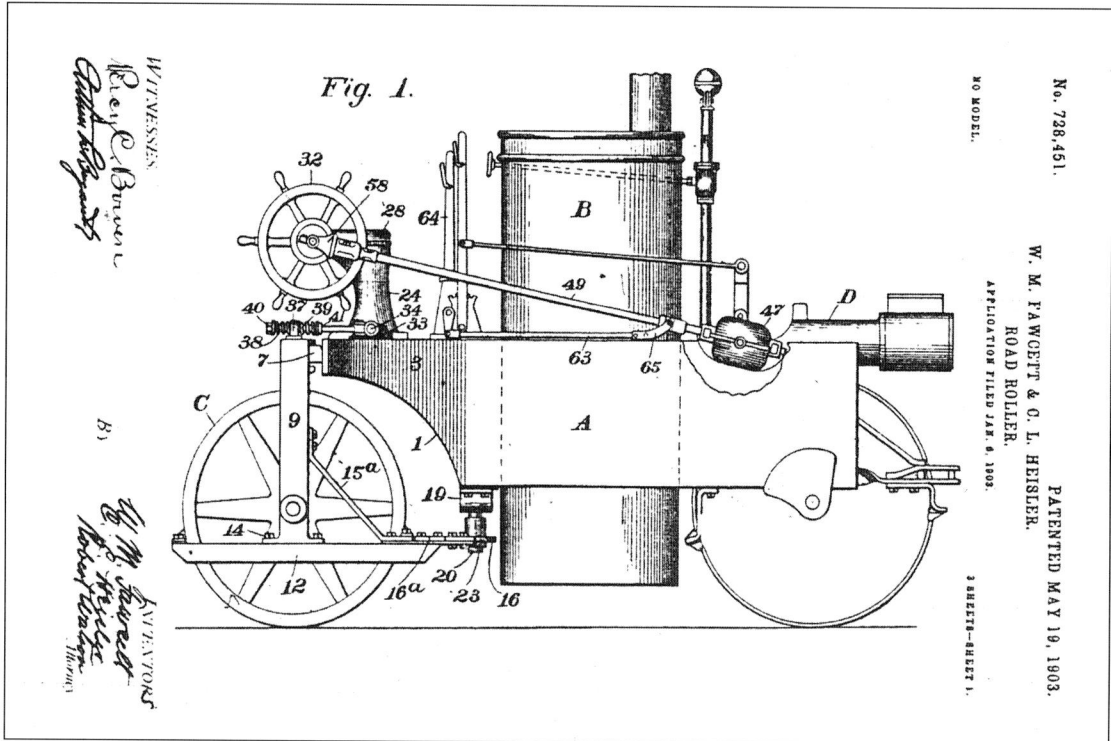

Here may be seen an original photograph of the unique Scholl roller. It is interesting to note that, while this machine shows the operator's position in the same place as in the patent drawing, there were other models of this roller where the engineer's seat and steering mechanism were located above the two cylinders.

On May 19, 1903, William M. Fawcett and Charles L. Heisler of Erie, Pennsylvania, received patent #728,451 for their tandem roller design, which they assigned to Julian Scholl and Company of New York City. Scholl built several of these steamrollers under the brand name of "The Universal Roller." He advertised them in the *Municipal Journal and Engineer*. Among the machine's peculiarities are the low-slung boiler, the tall frame, and the off-center smokestack.

THE BELL ENGINE AS A ROAD ROLLER.

This cut of an Imperial steamroller, also known as a Bell road roller, was built by the Robert Bell Engine and Thresher Company of Seaforth, Ontario. These machines were manufactured under license from the Port Huron Company of Port Huron, Michigan, and were an adaptation of Bell's farm traction engine. Like many Port Huron and Robert Bell engines, the roller has an enclosed cab that gave the engineer an unusual degree of comfort in inclement weather. The Imperial was offered in 10-, 12-, and 15-ton sizes. *Reynolds Museum*

Weight 4,600 lbs.

The regular driving wheels are taken off and heavy road roller wheels put on. This converting of a traction engine into a steam road roller is said to be one of the greatest steps towards solving problems as to bad country roads, because the attachments' comparatively small cost does away with the great objections to enormous investments of from $3,000 to $4,000 (much of the time idle) required for a regular steam roller.

This feature will also enable owners of our engines to put them to profitable use during the summer months when there is no threshing to be done, thereby greatly increasing their yearly earnings.

It is certain that owners of traction engines are to be to a large extent the builders of good roads in rural districts. Small towns and villages will also gladly avail themselves of the opportunity of being able to get this work done without investing in a steam road roller for their own use.

Diameter of Rear Wheels, 69 inches.   Diameter of Front Wheels, 40 inches.
Face of Rear Wheels, 22 inches.   Face of Front Wheels, 18 inches.

Grading with the Bell Engine.

This is Robert Bell's answer to Port Huron's "Combination" roller, or road locomotive. Chief among the differences between the two company's machines are Bell's single cylinder, as opposed to Port Huron's tandem compound cylinder, and Bell's sturdier front-wheel spokes, in comparison to the more gracefully designed Y-shaped spokes of the Port Huron front wheels. In company literature, Bell emphasized the weight of the wheels that transformed a traction engine into what the firm called a "road roller." The driver wheels weighed 4,800 pounds each. Bell saw its traction-engine conversion package as an inexpensive alternative for rural road builders and for small communities not wishing to spend money on machines designed to be steamrollers in the first place. *Reynolds Museum*

The Robert Bell Engine and Thresher Company, Seaforth, Ontario

Robert Bell first built sawmills and farm implements while running a repair shop in Hensall, Ontario. In the 1890s, he bought a boiler in London, Ontario. He and carpenter John Finlayson designed and poured patterns for a side crank engine, which they mounted on the boiler. This engine threshed in the 1899 season. Its success encouraged orders. Bell moved to a small foundry and factory in Seaforth. His business soon came to be known as the Robert Bell Engine and Thresher Company. Since traction engines were popular, Bell obtained permission to build Port Huron engines in Canada. In 1901, he constructed his first traction engine. In 1928, he built his last new engine, although rebuilds and repairs continued for a few years longer.

Bell's heavy roller wheels enabled traction engines to become steamrollers when not in use for threshing. Many Bell engines were sold through the Western Branch in Winnipeg and from distributors in Regina and Saskatoon. Bell built heavy threshers for the prairies and lighter machines for barn threshing in eastern Canada. Bell equipment was known as the Imperial line. In the early 1900s, the factory produced stationary, portable, and traction engines; steamrollers; boilers; separators; and sawmills.

117

Ross Road

The Ross Road Machine Company, Cincinnati, Ohio

Captain Abbott Q. Ross patented the first American roller on August 22, 1871. Steamboatman Ross's invention featured a vertical boiler, gear drive, two rolls for smoothing macadam roads, and a ramming device. Filling the main roll with water added weight to the roller and heating it melted tar. In 1873, Ross founded the Ross Road Machine Company to manufacture his steamroller. Eyewitness testimony confirms Ross also built three-wheel rollers with vertical boilers and chain drive. Dated July 1, 1940, Bert L. Baldwin's biographical sketch of Ross states: "In the early [1870s], Cincinnati claimed a good . . . engineer in Captain A. Q. Ross who was quite an inventor . . . the first large machine that he invented being the Ross Road Rammer, which was built on the general lines of the old Latta fire engine, with heavy frame mounted on three heavy, broad-faced cast-iron wheels, two in the rear and a single wheel in front arranged with a steering device. A vertical firebox boiler was mounted in the rear; the rear wheels were driven by a steam engine through sprocket wheels and chains. Back of the boiler and rear wheels was a stamp battery, consisting of double cams and five heavy tampers . . . The machine proved very efficient and satisfactory for rolling and tamping . . . macadam streets and roads. It has been claimed that this machine was the 'grandfather' of the present steam road rollers." The main drawback of this roller was that the rolls were rigidly mounted, causing the frame to bend when rolling uneven pavement. Ross is probably best known for inventing the stoker that bore his name. He died at age 62 in 1897.

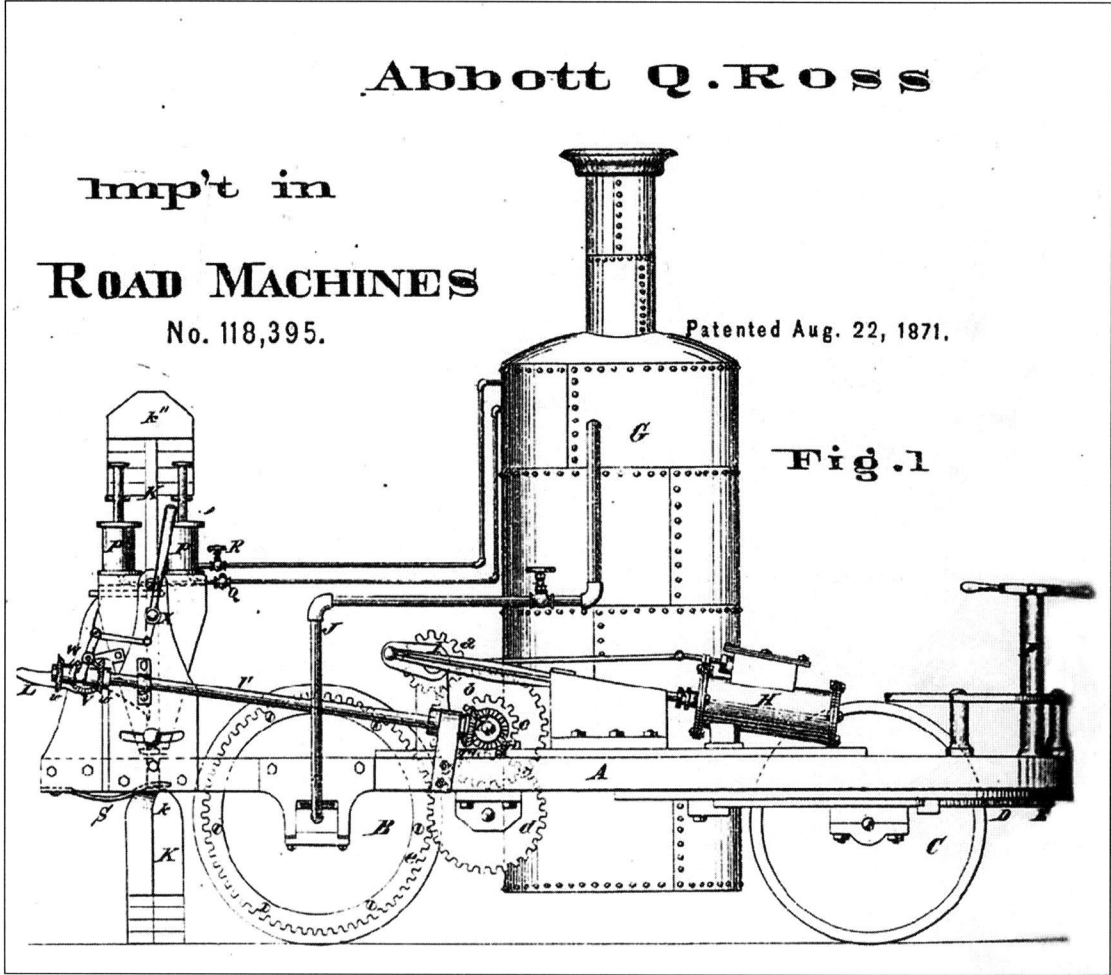

This machine represents the first patent issued to an American design for a steamroller; the patent was given to Abbott Q. Ross on August 22, 1871. Ross, who resided in Cincinnati, Ohio, not only invented the tandem roller but also designed the steam-operated tamping or compacting unit attached to the roller.

Here is a wonderful interior view of the final assembly area at the Massillon factory. Suspended from the overhead crane is a standard model roller, and at the right front is a rarely seen tandem compound steamroller. *Drake collection*

The Russell Company, Massillon, Ohio

In 1842, brothers Charles M., Nahum S., and Clement Russell, all carpenters, became partners in C. M. Russell & Company. They built houses, made furniture, and sold stoves. In the 1830s, Nahum and Charles had made "knock-outs," machines that knocked the grain loose from the stalk. By the mid-1840s, they were producing a thresher that included a separating/cleaning device. It was modeled on the Pitts brothers' design but included several innovations by Charles. In 1845, the firm built a steam engine to run the factory machinery. The Russell brothers also manufactured plows, mowers, and reapers. In 1853, Charles entered into partnership with Joseph Davenport and Marshall Wellman to build railroad cars. C. M. Russell died in 1860. In 1864, three more brothers joined the company: Joseph K., Thomas H., and George L. At this time, the firm name changed to Russell and Company. In 1871, brother Allen became a partner. The Russell Company was incorporated in 1878.

In 1883, the company welcomed the double-ported valve and the friction clutch designed by C. M. Giddings. The next year, Russell began using the Giddings reverse mechanism. By 1913, the buildings of the Russell plant covered twenty-six acres. Over time, the firm produced sawmills; threshers; horse powers; water tanks; stationary, portable, and traction engines; steam shovels; and steamrollers. Russell traction engines, road rollers, compound road locomotives, and steam shovels helped to build the road from Massillon to Canton to Alliance. A dummy locomotive ran on a narrow gauge railway parallel to the road, its small cars hauling construction materials.

Russell built its last steam engine in 1924; it was serial number 17152. After the company closed, a Mr. Schnider assembled four more engines from unused parts. At the time of this writing, George Kester has portable number 17155. In 1927, what remained of the company was sold at auction.

This picture is demonstrating the flexibility of the rocking joint on the Russell steamroller. The large flywheel that was typically installed on Russell steam engines can be seen clearly. The company stated in its advertising that this flywheel could also be used as a belting wheel. *Massillon Museum*

This picture from the 1907 Russell catalog shows an early steamroller built by that firm. Like most early rollers produced by manufacturers of farm traction engines, this roller has no belly tank, and the front steering yoke is made of riveted steel. *Massillon Museum*

Another picture from the 1907 Russell catalog, this one showing a cast-steel front yoke that, if nothing else, probably afforded easier access for cleaning the boiler tubes. This machine is fitted with a five-section front roll and has cleats on the rear wheels, indicating that it could also be employed as a road hauling engine. *Massillon Museum*

This picture from 1913 shows the evolution that the Russell roller underwent. Most noticeable is the addition of a belly tank, the massive front kingpin housing, and the transposition of the engine from the left side of the boiler to the right side. *Drake collection*

This 1913 right-hand view of the Russell roller shows that the firm apparently placed identification plaques on both sides of the kingpin housing. The toolbox mounted on the side of the coal bunker was made from cast steel. *Drake collection*

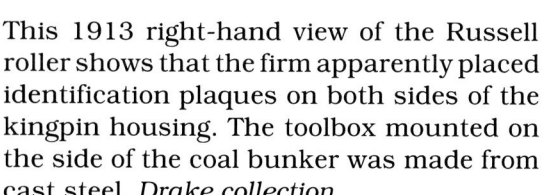

A right-hand view of the rarely encountered tandem compound steamroller. The Russell engine was unusual, in that the low-pressure cylinder was located at the front, while the high-pressure cylinder was at the rear. All Russell rollers came equipped with power steering and a brake, which is most unusual on steamrollers. *Drake collection*

A finished Russell steamroller that has been loaded on a railroad flatcar to be shipped to its new home in Brooklyn, New York. Worthy of note is the fact that Russell often placed advertising for the customer on the side curtains. *Drake collection*

The Waterous Company, Brantford, Ontario

In the early 1830s, Philip C. Van Brocklin, who had learned molding in New England, began working at a small iron foundry in Normandale in Ontario. There he met Elijah Leonard, also from New England. Brocklin and Leonard soon started their own foundry in nearby St. Thomas. Unfortunately, their business failed, and the partners went their separate ways. In 1841, Leonard established an ironware business in London, Ontario, and, by 1880, he was manufacturing boilers and steam engines. Van Brocklin moved to Brantford, where he built a foundry. His power source was a one-horse sweep in the basement. He manufactured plows, stoves, fireplace irons, and andirons.

Van Brocklin entered into several short-term partnerships, the first with Arunah Huntington, the second with F. P. Goold. The Goold partnership ended in 1848 when Van Brocklin rejected Goold's offer to buy him out. Goold began a competing foundry in Brantford. Van Brocklin's next partners were Thomas Winter and Charles H. Waterous. Having apprenticed in a machine shop in Brandon, Vermont, in 1834, Waterous had worked in Norwalk, Sandusky, and Cleveland, Ohio, and had sailed the lakes. By 1838, he served as chief engineer on the steamer *Governor Mansy*. He next moved to New York to join a former employer, Thomas Davenport. In the building where Davenport was designing a magnetic motor was Samuel Morse, who was developing an electric telegraph. The Davenport motor never provided enough power, and Waterous had no money left for further speculation. Waterous borrowed money to return to Sandusky, where he rented a machine shop. In 1839, he

married Martha June. Two years later, his business failed. He entered into a partnership with Julius Edgerton to manufacture grinding mills in Painesville, Ohio. In 1845, fire destroyed the factory. Waterous and Edgerton next formed a partnership with John D. Shepard to found the Shepard Iron Works in Buffalo. In 1848, a depressed economy forced the closure of the plant. It was then that Waterous arranged with Van Brocklin to take charge of the machine shop and foundry in Brantford. Waterous began producing sawmills and steam engines. In 1854, Waterous patented the first circular sawmill in Canada. By 1856, a 25-horsepower Waterous stationary engine replaced the tugboat engine that had replaced the horse. In the following year, Van Brocklin and Winters left the firm.

By 1864, the manufacturer was called C. H. Waterous and Company. In 1874, the firm incorporated as the Waterous Engine Works Company, Limited. During the 1860s, Waterous supplied engines and pumping machinery for the booming oil business.

David June of Fremont, Ohio, was the brother-in-law of Charles Waterous. Since 1877, D. June & Company had built a vertical-boiler engine known as the "Champion." June gave the Champion patent rights for Canada to Waterous, who began production in 1877. The firm eventually produced some 2,500 Champions. In 1881, Waterous provided a horse-steered, chain-drive traction engine. During the early 1880s, steam-powered fire-fighting equipment was added to the line. Waterous opened a branch in St. Paul, Minnesota, in 1888. In 1890, Waterous began mounting the Champion engine on a horizontal return-flue boiler. Waterous also produced the "New Economic Boiler," a return-flue

water-bottom firebox combination. In 1892, Charles Waterous died.

In the 1890s, the Waterous Company began building traction engines using the same gearing and controls as Buffalo Pitts engines. Some of the first of these were mounted on the "New Economic Boiler." The Pitts engines soon came in 12-, 14-, and 17-horsepower sizes as portables and in a 17-horsepower size as a traction engine. The boilers were the open-bottom locomotive type. By 1904, most of the firm's engines were double-cylinder models. Buffalo Steam Roller Company gearing also figured in the construction of Waterous steamrollers. Waterous road rollers came in 10-, 12-, and 15-ton sizes. By 1905, annual sales ranged from thirty to thirty-five steamrollers per year. The factory also turned out road graders and stone crushers. In 1910, the Waterous firm could not produce enough steamrollers to meet the demand and filled the quota by importing Pitts rollers. The firm began switching to gasoline engines and built its last steam traction engine in 1911, but it continued to produce steamrollers for several years.

If this Canadian made Waterous roller looks familiar to the reader, there is a good reason why. Waterous steamrollers were built at Brantford, Ontario, under license from the Buffalo Steam Roller Co. of New York. This machine is equipped with the early type headlamp which is attached to the kingpin cap. The pin stripe pattern seen on both the kingpin housing and front yoke is entirely different from U.S. machines.

In this left-hand view of this "Buffalo of the North," the reader can see that the roller is equipped with a belting wheel. All U. S. Buffalo and Buffalo-Springfield three-wheel rollers could be special ordered with belting wheels, except that the U.S. machines had a pressed steel wheel, whereas the Canadian models had a cast-iron wheel. *Reynolds Museum*

In this right rear view of the Waterous roller, several of the differences between it and its cousin from Buffalo, New York, may be seen. The scrapers for the rear wheels and the method by which they are mounted differ from the those used on U.S. machines. Waterous rollers were offered in 10-, 12-, and 15-ton sizes. This machine, with the counterweight cast into the flywheel, is either a 12- or a 15-ton model. The pinstripe pattern is different from that found on U.S. machines. *Reynolds Museum*

This rear view of the Waterous roller shows it fitted with twenty-four deep, or long, picks affixed to each of the driver wheels. These picks could be used either for deep compacting or to break up the surface of a hard-packed road to make the grading of that surface easier. Of note are the belting wheel to the left and the steering wheel, which differs from those found on U.S. machines. *Reynolds Museum*

MORE TITLES FROM ICONOGRAFIX:

*This product is sold under license from Mack Trucks, Inc. Mack is a registered Trademark of Mack Trucks, Inc. All rights reserved.

All Iconografix books are available from direct mail specialty book dealers and bookstores worldwide, or can be ordered from the publisher. For book trade and distribution information or to add your name to our mailing list and receive a **FREE CATALOG** contact:

Iconografix, PO Box 446, Hudson, Wisconsin, 54016   Telephone: (715) 381-9755, (800) 289-3504 (USA), Fax: (715) 381-9756

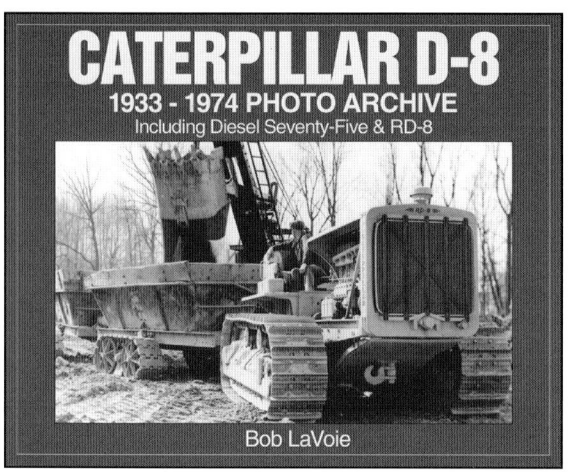

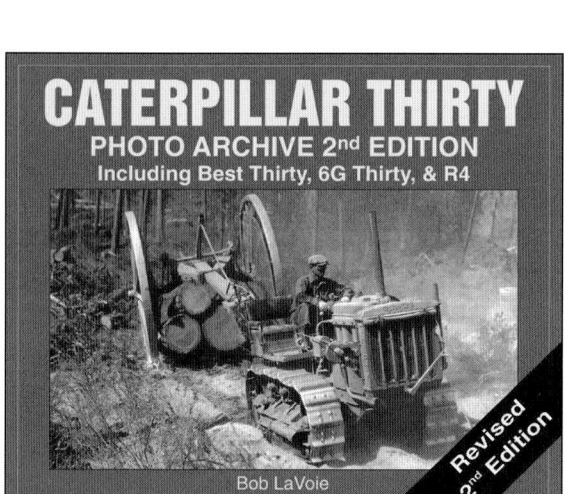